W9-BUO-655

The Practical Guide to Project Management Documentation

John J. Rakos, PMP

Karen Dhanraj
Laverne Fleck
James Harris
Steve Jackson
Scott Kennedy

WILEY
John Wiley & Sons, Inc.

This book is printed on acid-free paper. ♾

Published by John Wiley & Sons, Inc., Hoboken, New Jersey
Published simultaneously in Canada

For general information on our other products and services or for technical support, please contact our Customer Care Department within the United States at (800) 762-2974, outside the United States at (317) 572-3993 or fax (317) 572-4002.

Wiley also publishes its books in a variety of electronic formats. Some content that appears in print may not be available in electronic books. For more information about Wiley products, visit our web site at www.wiley.com.

Library of Congress Cataloging-in-Publication Data:

Rakos, John J., 1946–
The practical guide to project management documentation / John J. Rakos, with Karen Dhanraj . . . [et al.].
p. cm.
Includes bibliographical references and index.
ISBN 0-471-69309-X (cloth)
1. Project management. I. Dhanraj, Karen. II. Title.
T56.8.R35 2004
658.4′04—dc22

2004009415

Printed in the United States of America

10 9 8 7 6 5

This book is dedicated to my wife, Marie, who always helps me write but rarely gets credit for it.

John Rakos

Contents

Preface

Purpose of This Book

Most of the deliverables in the project management process are documents. Therefore, it is imperative to produce these documents correctly.

Many books have been written on project management. All those books refer to project documents, and most of them outline the major documents required to produce a project. However, those books rarely include detailed examples of those documents, and we have not yet seen one in which electronic versions of the documents are available to use as templates.

This book therefore will help the reader produce excellent project documentation by providing descriptions and detailed examples of project management documents.

Nonpurpose of This Book

It is not our intent to teach the reader project management. There are many excellent texts in this field (do a search on the Web for "Project Management," starting with www.pmibookstore.org). In fact, the reader must be familiar with project management basics to write the documents detailed here.

Origins of This Book

The primary author, Professor John Rakos, teaches project management to the Master of Business Administration class at the University of Ottawa. This twelve-week class covers every aspect of project management. The students have twelve group assignments to complete, consisting mainly of project documents. One of the groups in a recent class did such an excellent job of producing those documents that we thought it worthwhile to publish them. The authors of this book are therefore John Rakos, assisted by MBA students Karen Dhanraj, Laverne Fleck, Jim Harris, Steven Jackson, and Scott Kennedy.

Organization

The book is divided into four parts based on the major project phases defined by the Project Management Institute's *A Guide to the Project Management Body of Knowledge* (*PMBOK Guide*). These phases are Initiation, Planning, Execution and Control, and Closing. Each phase produces one or more documents. Each section of the book is subdivided into chapters, with one chapter dedicated to each document type. Therefore, these documents appear in the same chronological order in which they are produced during the implementation of the project.

For each type of document there are two parts: discussion and example. In the discussion we explain the purpose, author, timing, and other attributes of the document. We give the outline of each section of the document and explain the content, pointing out any problem areas to beware of. We then present an actual detailed example of the document.

The Case Study Project

The documents are based on an actual project: the development of a water theme amusement park to be constructed in Ottawa, Canada (see Figure P.1). Although the documents are true to the concept of managing the construction of the water park, the actual players, figures, estimates, events, and problems are fictitious.

To better understand the examples, the following information is offered. Ottawa is the capital of Canada, located in the province of Ontario. Ottawa–Carleton is the former name of the regional municipality managing all of Ottawa and its surrounding suburbs. Nepean (pronounced "Nipea-an") is a suburb on the west side of Ottawa. The National Capital Commission (NCC) is a government body with broad powers over the types of development that are permitted in the Ottawa area. KLSJ Consulting is a fictitious firm created by the students (*K*aren, *L*averne, *S*teve, *S*cott, and *J*im), and Carlington Aquatic Parks (CAP) is the actual company created by the clients who had the original idea to build a water park.

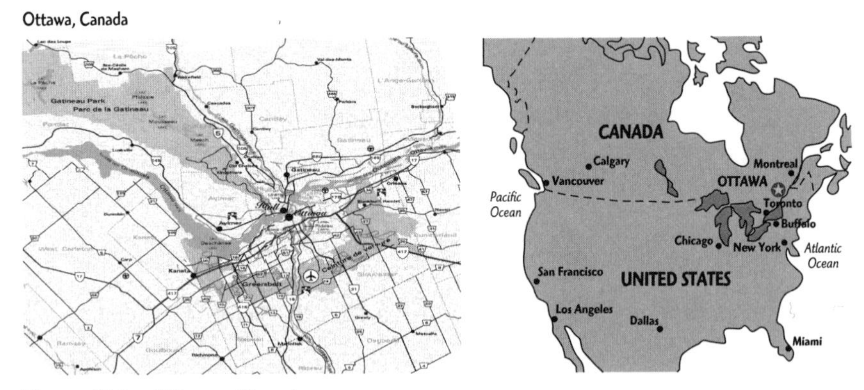

Figure P.1 Ottawa, Canada.

In this fictitious scenario, the project was started with CAP developing the Project Concept and Project Charter in late 2002, followed by a bidding process for a prime contractor. The contract was won by KLSJ, which then was hired to deliver a completed water park to CAP by spring 2005. Most of the example documents are ones that KLSJ would have had to create either for the client or for internal management.

Who Are You?

This book is intended for the project manager, project leader, team leader, sponsor, client, or even end user of a project—anyone who must produce or read project documentation. Obviously, the book can be used as a text for any course that teaches how to write project documents. We attempted to be simple, clear, and concise; the documents have been pared down to the essential information.

To the Teacher

This book is intended as both a management text and a teaching text. It will prove to be an invaluable tool for learning or developing project management skills and for training in a school or internal training environment.

CD-ROM

The CD at the back of the book contains all the document examples in the form of Microsoft Word text files. The reader is welcome to use these files as a framework to develop his or her document. Normal copyright rules apply, however: If you use the texts verbatim, reference to this book must be made.

Services from the Author

John Rakos has an Internet site at www.rakos.com. Updates and improvements to the templates in this book will be made available at this site. Additional templates for other types of projects will be made available as well. He can also assist in customizing these documents for specific projects and applications. Mr. Rakos can provide consulting help in almost any aspect of project management and also can teach seminars and lectures on project management topics. See www.rakos.com for details on the services provided.

Contacting the Authors

The authors can be contacted at the following addresses: John Rakos (john@rakos.com), Karen Dhanraj (karen.dhanraj@vopakcanada.com), Laverne Fleck (karlav@sympatico.ca), Jim Harris (jrharris@storm.ca), Steven Jackson (gygsrj@magma.ca), and Scott Kennedy (scott.wendy@rogers.com).

Acknowledgments

First, I would like to acknowledge the comments and encouragement of Professor Gilles Pacquet, former Director, Center on Governance, University of Ottawa. Second, I would like to thank Mr. Dan Milks, President and CEO of Carlington Aquatic Parks Ltd., for his permission to let us borrow his idea for a water park, first as a project for the MBA project management course and then as the basis for the examples in this book.

Introduction

The Importance of Documents

Until a project produces some tangible deliverables, the only items produced are the documents. In fact, documents are considered some the major milestones of the Initiation and Planning phases. They drive the project, organize it, standardize it, and provide communication not only among the stakeholders but within the project team. If one member of the team leaves, the replacement should be able to take over strictly by reading the documents. If the team in one phase is not the same as that in another, documents are the main form of information exchange. Imagine what happens if the planning team does not produce a proper Project Plan. Can anyone continue working on the project without a clear idea of what is to be produced, for whom, and by what date? Some of you may be saying "I've been there!" but it is better to write it down.

In a time crunch, the first item to be dropped is the production of the documents, but a wise project manager knows that the documents are the first items to produce. Without them, the project flounders, wrong things are built, no one is aware of progress, and the project dies.

Standards

In a large organization in which many projects are produced, one of the most efficient project management methods is to have all the documents with a similar look and feel. You can accomplish this by using the outlines and templates provided in this book. Most important, having standard documents will introduce standard terminology, which will improve communication among project stakeholders immensely.

The Documentation Plan

Why Have a Plan?

Although this book is not intended to teach project management, the basic tenet of project management is to plan things. Since documentation will be the only deliverable for the first part of the project—and in fact sometimes the only tangible deliverable (in the case of a software project, for example)—it has to be done right. This book will help you plan.

Size and Time

Make your documents as concise as possible. It is very important to keep the project's size in mind when you decide what to report. For example, even for very large projects, clients will not pore through hundreds of pages of minutiae to determine the status of their project. It therefore is best to group comments into shorter statements based on activity or outcome, with supporting documentation provided in appendixes for lower-level managers to review. You can write only five to ten pages a day, so use your time efficiently.

Who Is the Document Intended for?

There are always obvious and less obvious readers for each document. For example, the Project Concept is intended for the financial decision makers, and the Project Plan for the project approvers and workers. However, documents are public: You do not know who may be reading them. I always recommend writing for the lowest common denominator. This can be a high-level manager who is unfamiliar with project management and its terminology. However, the project may depend on the approval of this manager, so write the document to ensure that he or she understands it.

Language

Keep the language simple; include only the essential information. There is no need to impress anyone with large, complex words and sentences. KISS (*K*eep *I*t *S*imple, *S*tupid) applies here as it does in any other writing.

The Case Study Project

The documents in this book are based on an actual project: the construction of a water theme amusement park in Ottawa, Canada. Although true to the theme, the actual problems, issues, and concerns raised in the example documentation are fictitious. However, they illustrate typical problems that can arise during any project.

The Order and Necessity of the Documents

Table 1 lists all the documents detailed in this book. The table is divided horizontally for two types of projects: internal and external. An internal project is one produced by one section of a company for another section. There is no formal contract between the developers and the client. An ex-

Table 1: Documents Produced by Phase of the Project

Phase		***Initiation***		***Planning***				***Execution and Control***	***Closing***
Internal Project	Project Concept	Business Case	Project Charter	Preliminary Plan			Final Plan	Status Reports	Postproject Report
							Communication Plan		
							Risk Management Plan	Risk Control Report	
							Quality Management Plan	Quality Assurance/Quality Control Report	
								Meeting Minutes	
External Project	Project Concept (by Client)	Business Case (by Client)	Request for Proposal (by Client)	Preliminary Plan	Proposal and Evaluation	Contract	Same as Internal Project	Same as Internal Project	Same as Internal Project

ternal project is one contracted with an external developing organization and involves a formal procurement process and contracting.

Not all of the documents in this book are produced for every project. Let us detail the circumstances under which each one would be produced (see Table 1).

Project Concept

This document must be produced for every project. It is not a large effort (two or three pages), outlining the basic ideas, problems to be solved, strategy, and solution, plus a ballpark cost and time estimate. The estimates in this document may be +75% to −25%* in error.

The Business Case

This document must be produced for every project. It is used to ensure that the development effort is cost-effective; the project must eventually pay for itself.

Requirements (Internal Projects) or Request for Proposal (RFP) (External Projects)

The requirements document details the client's business problems that will be solved by an internal group. It states the maximum budget allotted, the required time frame, and possibly a suggested solution. Depending on the knowledge the client has of the project requirements and the skill of the person writing it, this document may be as short as a few pages, giving only rudimentary requirements, or as long and as detailed as the Preliminary Plan.

The RFP accomplishes the same purpose for external projects. This is the document published to the external world, soliciting the interest of contractors to bid. Some items that may appear only in the RFP are a request for data on the contractor (such as experience and references) and the more formal terms and conditions of the solicitation.

Preliminary Plan

This is the first, high-level plan of the scope, time, cost, communication, risk, quality, procurement, and human resources required. For an external project, this is the basis of the proposal; for an internal one, it is the basis of the Project Charter. The estimates in this plan may be +25% to −15%* in error.

---

*Max Wideman, http://www.maxwideman.com/issacons3/iac1332/tsld006.htm.

The Proposal (External Projects) and the Charter (Internal Projects)

For the external and internal worlds, this is the formal statement of the developers to the client about the exact deliverables, cost, schedule, method of delivery, acceptance, commitment, and so forth. In a competitive environment, the Proposal is also a sales tool, emphasizing the virtues of the vendor. Obviously, the external Proposal leads to a more formal contract than the internal Charter; however, a formal commitment should be made in both cases.

Contract

For larger external projects, a contract may be needed for legal purposes. For small projects, a signature on the proposal may suffice.

Final Plan

It may take several weeks or possibly months to go through the contracting process, and more information about the project will then be available. In addition, the estimates in the preliminary plan might have been very inaccurate. The preliminary plan therefore is revised, more detail is filled in, and the estimates are redone to generate the final plan.

Communication, Risk Management, and Quality Plans

The ways these items will be managed must be addressed for all projects. Depending on the size and scope of the project, these may be separate plans or sections in the final plan. Examples of stand-alone plans are discussed in the book.

Meeting Minutes

Two meetings are discussed: the team status meeting and the Project Managers meeting. The former is informal, and minutes may not be taken. The book shows a "record of discussion" with a list of action items. The Project Managers meeting occurs in most organizations in which a committee of managers oversees the progress of several projects. Formal minutes should be taken, and this is shown in the book.

Status Reports

When the project execution phase starts, progress status must be reported to all the stakeholders at a set frequency. A simple status report is shown in the book.

Risk and Quality Control Reports

These items have to be monitored constantly, and changes, issues, and status must be reported to stakeholders. Again, depending on the size and scope of the project, these may be separate reports or sections of the status report. Examples of stand-alone reports are discussed in the book.

Subcontract Request for a Proposal, Proposal, and (Sub)Contract

Note that the flow of events assumes that the whole project may have been contracted (see the external project documents in Table 1). When portions of the project are subcontracted out, another procurement process may occur. In this case another set of documents—the Subcontract Request for a Proposal, Proposal, and (Sub)Contract—may have to be written. For accuracy, we repeat the titles of the documents but do not repeat the discussions; we simply refer the reader to the appropriate documents.

Post-Project Report

This is a stand-alone document that details how the project started, how it was run, what went well, and what did not. It is a crucial "lessons learned" document that all projects must produce.

Items *Not* Included in the Examples

The following topics could be included in *every* document but were omitted for the sake of brevity. Consider including the following topics for a large, highly visible project in a formal environment.

Document Change Control

A formal versioning of the documents can be implemented by prefacing the text with a paragraph identifying the date, author, purpose, and description of the change.

Glossary

If technical jargon or acronyms must be used, the document can end with a glossary.

Associated Documents

If other documents are referenced, a paragraph listing the location of those documents can be included.

Initiation Phase Documents

Project Concept

Project Concept: Discussion

The Project Concept is the initiating document for the entire project. It is intended to be reviewed and approved by the client and is a foundation document for the remainder of the project. It defines the business requirement or *need* for the project in a broad sense and is used in making a preliminary go/no-go decision.

Activities that can occur in the preparation of the Project Concept include concept exploration, feasibility studies, demonstrations, and a proof of concept. The document should be no more than one or two pages in length, although additional attachments such as maps, drawings, photographs, and other types of illustrations may be used to engage the client's interest.

In addition to being used for the decision to proceed to the planning phase, the Project Concept is helpful in categorizing the work to be done, setting initial priorities, and obtaining preliminary funding approval. The Project Concept may be used to obtain the funding for the whole project or, more likely (and more wisely), the funding for the planning phase. In a large organization, the decisions based on Project Concept documents may set the direction for the organization.

Project Concept Outline

Executive Summary

This is an optional section that summarizes the major points of the document. It gives readers a chance to decide whether they need to read the remainder. However, because of the typical brevity of the Project Concept document, an executive summary is probably unnecessary.

Background

Provide the readers with a short summary of the situation, including project location, the scope of the project (local, national, or international), and any history or biographical information that would provide context to the reader. Remember that readers may not know anything about the project.

Challenge

In this section, describe the compelling "reason for being" for the project. Lay out the challenge in a step-by-step argument. Define the challenge: what it is, the cause, when it occurs, where, and how much it costs. Without exaggerating the facts, leave the reader with a clear sense of a need for something to be done. If there is more than one possible interpretation of the challenge, discuss each one in turn and examine any alternative sources or causes as well as possible courses of action. If possible, state the cost of *not* doing anything about the problem.

Suggested Solution

Describe the proposed solution in a clear, unequivocal statement. List the major deliverables to be produced, both product and process, by whom, for whom, what, when, and where. Give a ballpark cost, a duration, and a delivery date. Do not worry about accuracy here: This estimate may be +75% to −25% in error.

Describe the general approach or strategy for managing the project. If there are several alternative solutions, describe each and suggest one. Some alternatives would be to build it ourselves, contract it out, and have different parts done by different groups. The project concept tries to identify different approaches for efficiency; for example, release a prototype in one region, and then roll it out to others after acceptance. Each alternative is evaluated in terms of price, schedule, and meeting the client's constraints of time, money, and quality. One alternative that must be evaluated is to "do nothing," that is, the cost of not doing the project.

Project Concept: Example

KLSJ Consulting

September 7, 2002

Project Concept

Ottawa–Carleton

Water Park

14 Palsen St., Ottawa, ON,
Canada, K2G 2V8

Background

Ottawa is the nation's capital and the fifth largest city in Canada, with a regional population of over 1 million. As a summer tourist destination, Ottawa ranks third nationally, with approximately 5,500,000 visitors, of whom roughly 2,000,000 visit during the summer. Ottawa has some of the finest exhibits, cultural expositions, galleries, and recreational facilities in Canada. It also has one of the highest average disposable incomes in North America.

The Ottawa–Carleton region is growing rapidly, with projected annual population growth of over 5% throughout the next decade. Other than the federal government, the largest business sector is high technology, which has enjoyed incredible success recently and expects that trend to continue. Private land is becoming relatively scarce and expensive. However, there is an opportunity to petition for the use of public land set aside specifically for recreational development.

There are over 1400 water parks in over 100 countries around the world, with over 900 in the United States alone. It is a mature industry with a good track record for profitability and a central association that offers considerable support and guidance to new operations. Water parks typically operate from late May to early September, with little variation between northerly and southerly locations.

Challenge

With all the attractions and facilities in the Ottawa area, one key recreational destination that is lacking is a full-size water park. Preliminary demographics point to the potential for a highly profitable water park facility in Ottawa. Financial projections show that after the first year, profits in excess of $1.5 million annually can be expected.

An average-size water park typically includes towers, slides, a river, a wave pool, a children's center, an adult spa, and a group area with picnic facilities. Other attractions, such as sports courts, a rock-climbing wall, and a mini-putt and driving range, typically round out the facility and provide good cross-usage potential. Although all or most of these facilities exist somewhere in the Ottawa area, there is no single site that has all these attractions.

With the continuing strong popularity of water parks across North America, it will not be long before a firm steps forward to exploit the Ottawa market. First-mover advantage would convey almost monopolistic rights to the local market, since Ottawa is not large enough to support more than one water park profitably. There is a need therefore for rapid action to cultivate financial support, plan and develop the site, and open the facility before another company is able to do so.

Suggested Solution

It is proposed that a full-size water park be built on a suitably accessible site in the Ottawa area, with an initial estimated investment of Can$12 million. Ideally, this site will be leased from National Capital Commission public lands, with development costs shared by the City of Ottawa. The facility will include all the attractions mentioned above and will be built as a turnkey operation by a prime contractor to be chosen through a competitive bidding process. The owners and investors will act as a supervising agency and will retain authority for the approval of all creative and business decisions related to the project.

The project, which is expected to take almost two years to complete, should begin no later than the summer of 2003 and be completed by the May 21 weekend, 2005.

2

Business Case

Business Case: Discussion

Once you have an idea and have articulated it in a general form in the Project Concept, it is time to assess how good the idea is. In other words, is there value in proceeding with the project? This assessment is done by means of a business case.

A business case is a *logical*, *objective*, and *comprehensive* approach to analyzing the issues surrounding an initiative or a potential project. Its purpose is to support the decision-making process, in effect, to help decide if there is sufficient benefit to invest time and resources in the project. The steps in the development of a business case must be laid out clearly and easy to follow (logical). The Business Case must provide a balanced assessment of the pros and cons of the initiative (objective), and it must provide a thorough discussion of the issues from a quantitative as well as a qualitative perspective (comprehensive).

A business case builds on the Project Concept and helps flesh out the initial concept by putting meat on the bones. It facilitates early stakeholder engagement and consensus in the decision process by providing managers with a precise and disciplined medium of communication. Most important, a business case produces recommendations that are transparent, concrete, and defensible. Assumptions are stated clearly, and project risks are identified and factored into the decision-making process.

A business case usually is developed by one of two individuals. In many cases, the person who has the idea in the first place—the sponsor or client—does it. He or she wants to be certain that the project is viable and that the risks are well understood before continuing. However, for an internal project, responsibility for the development of the Business Case may fall on the shoulders of the project manager and his or her team.

It is important to note that a business case is not a full-blown "business plan." The latter has a broader focus and is much more of a management-planning document that frequently is aimed at obtaining investment funding. It normally includes a detailed assessment of the potential market, sales and promotion plans, an outline of the concept of operations after the project has been delivered, and a detailed financial assessment. The Business Case, while it provides the building material for the business plan, is more limited in its goal of facilitating good decision making at the front end of the project.

Just as there are an endless variety of projects, there are many forms and formats to a business case: One size does not fit all. A small, straightforward project may require only a couple of pages to argue its worth, and the Business Case could be pro forma. In contrast, a large, complex project may require a very sophisticated cost-benefit analysis employing probabilistic risk analysis that could stretch to several hundred pages or more. The depth of research needed to acquire the data to develop a business case also varies substantially from a few days for a simple project to many months for a highly complex one.

Some projects require a simple go/no-go decision. In other situations the decision to proceed has already been made; for example, government legislation on environmental protection may require that some action be taken by a certain date. The decision then becomes one of choosing among various possible options. In either situation, the basic purpose of the Business Case does not change: It must support the decision-making process involving whether or not to proceed with the project and, if the decision is to proceed, the preferred option to take.

Business Case Outline

Executive Summary

This is an optional section that summarizes the key points of the Business Case. It should provide enough detail that the reader has a clear understanding of the remainder of the document and can decide if further reading is required. At a minimum, this section should highlight the key factors behind the cost-benefit analysis, the results of the analysis, the risks involved in the project, and, most important, the conclusion as to whether the project is worth undertaking. Normally, an executive summary is limited to one or two pages, and it may not be required in a short, straightforward business case.

Aim of the Project

This should be a short, clear statement of the aim of the project, usually not more than a couple of sentences.

Background

The purpose of this section is to provide a short summary, perhaps along with a bit of history, of the intended project, including its scope, where it is to occur, and when. The aim here is to provide context to the reader. Much of this background can be taken from the Project Concept.

Key Assumptions and Constraints

The main assumptions in building the Business Case need to be stated clearly. This is key to providing understanding of and transparency into the way the Business Case was developed. Also, it is useful to describe any constraints or limitations that were encountered in conducting the analysis, such as limited data in a particular area.

Cost-Benefit Analysis

The cost-benefit analysis is the heart of the Business Case. It must be carried out in a clear, easy to follow, and objective manner. The natural tendency to be overly optimistic must be avoided; after all, your reputation or money may be at stake. As this can be a lengthy section of the document, it is best to divide it into several parts. It may be useful to put lengthy tables of figures and calculations into appendixes to keep the main body of the document relatively concise. A simple division is as follows:

1. *Revenue (or Benefit) Analysis.* Calculate the potential cash flow or revenue that can be earned from the project. Be sure to indicate how the data on revenues were obtained and any assumptions that were used in the calculations. This frequently requires discussions with sales managers and marketing research personnel. Without question, the estimation of potential revenues is the most difficult part of the business case equation. If possible, give an estimate of the degree of accuracy of the numbers. This sometimes is expressed as a probability; for example, mean estimate is Can$1,000,000 per year*, with a 20% probability of less than $800,000 and a 10% probability of more than $1,200,000.
2. *Cost Analysis.* As with the previous section, calculate the estimated costs of the project. This analysis needs to include the costs to develop the product as well as the estimated annual recurring costs to market and sell the products developed. In the

*All dollar figures used throughout the book are in Canadian dollars. Occasionally, we will designate Can$, to remind readers of this fact.

case of the water park, the annual operating costs have to be calculated. Preferably, these costs should be broken out into fixed (costs that do not vary with the amount of product produced or used) and variable costs (costs that vary directly with the amount of product produced or used). An accurate cost analysis is usually much more easily developed than is the revenue estimate, the exception being a project with an extensive research and development (R&D) component.

3. *Project Evaluation.* Essentially, this is a comparison of revenues (cash flow) to costs to determine if it is worthwhile to undertake the project. Several methods can be used to make this comparison.* These methods also can be used to determine the preferred option in a project. The three most commonly used project evaluation methodologies are:
   - *Payback Period.* This is the simplest method and is defined as the expected number of years required to recover the original investment. All that is required is to divide initial cost (investment) by anticipated annual profits. Although simple, this method has a serious limitation in that it does not consider the time value of money and thus is not recommended.
   - *Internal Rate of Return (IRR).* Companies (and individual investors) most often have a desired rate of return they expect from an investment. By calculating the rate of return from the project at hand and comparing it to the desired rate, one can determine if the project is worthwhile. The IRR is superior to the payback period methodology in that it takes into consideration the future value of money, but it can be difficult to calculate for irregular cash flows.
   - *Net Present Value (NPV).* Conceptually, this is the best method to use. It discounts all cash flows back to present values at the project's cost of capital and then sums those cash flows. The project is accepted as worthwhile if this sum, the NPV, is greater than zero. This method thus takes into account the future value of money and can handle any type of cash flow.
4. *Sensitivity Analysis.* It is often very useful to do a sensitivity analysis on one or more of the key estimates. This is done by systematically varying one of the parameters in the calculations to determine its effect on the final outcome. If it is found that the ultimate decision is highly sensitive to a particular estimate, it may be worthwhile to devote additional time and effort to re-

*Most good financial management textbooks contain a description of the three methodologies outlined here, as well as others, along with the advantages and limitations of each. Anyone undertaking a business case analysis (financial books usually call it a capital investment analysis) would be well advised to consult such a reference.

fining that number. At the very least, you will be aware that this estimate has risk associated with it. One form of sensitivity analysis is determining a break-even point, usually with respect to sales. If the break-even figure is clearly within reach, it provides a sense of comfort that the project will not lose money even if it may not have the desired rate of return on investment.

Risk Assessment

At this stage of the project it is unlikely that there will be a full understanding of all the risks involved. Nevertheless, a first step in this regard should be taken. The key risks, at a macro level, should be identified, with brief comments provided on each one. A full risk assessment will need to be undertaken at a later date in the project (see Chapter 7). Normally, a subjective assessment is made as to whether the risks are acceptable in light of the expected value of the project.

In some cases, particularly if the final decision hinges on a comparison of the different risk levels with the costs and profits in several options, it may be necessary to factor risk more objectively into the decision making. This can be the case if the decision is between a medium-cost but high-risk option and one with high cost but low risk. This type of analysis can be done by quantifying each risk as to both likelihood and impact and then running a computer simulation to determine a probabilistic cost-benefit analysis. However, this is a sophisticated methodology, that requires specialized knowledge and skills beyond the capabilities of most project teams. External assistance normally would be required, but this may be money well spent.

Conclusion

It is now time to wrap up. Clearly state if it is worth proceeding with the project and why. It is also important to reemphasize any major risk areas affecting the project. Keep this section of the analysis short and to the point.

Business Case: Example

KLSJ Consulting

September 30, 2002

Business Case

Ottawa–Carleton

Water Park

14 Palsen St., Ottawa, ON, Canada, K2G 2V8

Executive Summary

Project Concept

Ottawa is a rapidly growing city with a regional population of over 1 million and many excellent attractions and recreational facilities. The city continues to experience strong economic growth supported largely by the high-technology community. Ottawa is also a major tourist destination with approximately 5,500,000 visitors attracted to the city each year, of whom an estimated 2,000,000 come in the summer. However, the region is lacking a key recreational attraction: a full-size outdoor water park.

Water parks are a mature industry with a proven record of profitability. A typical water park includes towers, slides, a river, a wave pool, a children's center, and an adult spa, as well as attractions such as a rock-climbing wall and a mini-putt range. The location of such a park is critical, with the ideal site being within a major urban area to minimize transportation times.

The aim of this project is to develop a full-size water park in the Ottawa-Carleton region with an opening date in May 2005.

Key Assumptions

The following key assumptions were made in developing this Business Case:

- A suitable site can be leased from the National Capital Commission lands.
- Suitable financing of 50% debt and 50% equity can be arranged.
- Inflation will remain at an average annual rate of 2.5%.
- The Water Park would be built by a prime contractor and then turned over to an experienced water park management company for ongoing operations.
- The park would operate from the end of May to early September each year (a 15-week period).
- Financial calculations are based on an operation period of 25 years.

Cost-Benefit Analysis

Revenue Projections

Drawing from World Waterpark Association databases and applying them to Ottawa–Carleton demographics, an estimated annual attendance of 308,000 visitors to the Water Park was derived. Such a number would require a site with a capacity of 4500 guests with room for expansion. Based on competitive pricing, estimated annual revenues are:

Gate receipts	$5,729,000
Parking fees	$289,000
Concessions	$1,540,000
Total	$7,558,000

Cost Analysis

Initial cost figures for a water park of this size are based primarily on industry data obtained from the World Waterpark Association. Estimated costs are:

Construction of Water Park	$12,000,000	(4,500-visitor capacity)
Future refreshing of attractions	$1,000,000	(every three years)
Annual operating costs	$5,066,000	(30% fixed, 70% variable)

Evaluation

Net Present Value (NPV) methodology was used to determine the financial viability of the venture. In applying this method, a nominal cost of financing of 9% (6.5% return over inflation of 2.5%) was assumed. The NPV result was a positive Can$6,020,000, indicating a sound financial undertaking with a very good rate of return.

Sensitivity Analysis

Sensitivity analysis determined that the break-even attendance was 240,000 visitors annually, a 22% reduction from the estimated baseline calculation of 308,000. As the baseline figure has an estimated accuracy of ±20%, there is high confidence that the break-even attendance figure can be surpassed.

Similar sensitivity analysis was used to determine that the break-even cost of financing was a real rate of 11.5% (nominal rate of 14%). Looked at another way, if the baseline projections prove valid, the project will have a nominal rate of return of 14%—a very healthy rate.

Risk Assessment

There are several risks inherent in this project. Broken out by phase, these risks include the following:

Planning, Construction, and Set-to-Work Phase

1. Obtaining financing is considered a major hurdle because of the current competition for investment funding. This is the highest-risk area in the initial stages of the project.
2. Construction delays would jeopardize a May 2005 opening. Given good project management, this is considered medium-risk.
3. Although construction cost overruns are common in most large projects, good project management can control them. This risk is considered low to medium.

Operating Phase

1. Estimating annual attendance is a key element in forecasting the financial viability of the Water Park. However, analysis indicates that there is a very high likelihood of achieving the break-even attendance of 240,000 visitors. This risk is therefore assessed as low to medium.
2. Weather is a major determinant in the viability of a water park. Analysis indicates that the Ottawa area has weather that is very conducive to such an attraction and that even in a poor-weather year, the park can expect to at least break even.

Recommendation

Given that a good rate of return is anticipated with no undue risk, it is recommended to proceed with this project.

Aim

The aim of the project is to develop a full-size recreational water park in the Ottawa–Carleton region with an opening date in May 2005.

Background

Ottawa is the nation's capital and the fifth largest city in Canada, with a regional population of over 1 million. The city has some of the finest museums, cultural expositions, galleries, and recreational facilities in Canada, attracting approximately 5.5 million visitors each year. With all the attractions and facilities in the Ottawa area, a key recreational destination that is lacking is a full-size outdoor water park.

Preliminary demographics indicate the potential for a highly profitable water park facility in Ottawa. The Ottawa–Carleton region is growing rapidly, with projected annual population growth of over 5% throughout the next decade. Other than the federal government, the largest business sector is high technology, which has enjoyed incredible success recently; the expectation is for that trend to continue. The city also has one of the highest average disposable incomes in North America. Although private land is becoming relatively scarce and expensive, there is an opportunity to obtain the use of public land set aside specifically for recreational development.

With over 1400 water parks in 100 countries around the world and more than 900 in the United States alone, water parks are a mature industry with a good track record for profitability. An average-size water park typically includes towers, slides, a river, a wave pool, a children's center, an adult spa, and a group area with picnic facilities. Other attractions, such as sports courts, a rock-climbing wall, and a mini-putt and driving range, can round out the facility and provide good cross-usage potential. Although most of these facilities exist individually somewhere in the Ottawa area, no single site has all these attractions.

The project, which is expected to take over two years to complete, should commence detailed planning in the fall of 2002 and be ready for opening by the end of May 2005.

Key Assumptions and Constraints

The following key planning assumptions are made:

1. A suitably accessible site in the Ottawa area can be leased from the National Capital Commission public lands. Preliminary consultations indicate that such an arrangement is possible.

2. Suitable financing can be obtained. Discussions with financial institutions indicate that they would provide loans covering up to 50% of the initial investment provided that the remaining 50% was equity-financed.
3. Inflation will remain at an average annual rate of 2.5% (the midrange of the Bank of Canada's long-term target).
4. The Water Park would be built as a turnkey operation by an experienced prime contractor and then turned over to a water park management company for ongoing operations.
5. The Water Park would operate from late May to early September, typical of water parks in North America.
6. The business case is based on an operational period of 25 years, with periodic recapitalization.

Most of the revenue and cost data were obtained through the World Waterpark Association, which has over 1000 members active in all facets of the industry. The association maintains a statistical database covering all aspects of operations and offers support and guidance to new operations. Therefore, although there is no similar facility in the Ottawa area on which to base projections, the figures obtained from the association have a high degree of relevance for the local marketplace.

Cost-Benefit Analysis

Revenue Analysis

Attendance Calculation

The Ottawa–Carleton regional population within 40 kilometers (26 miles) of the water park is estimated at 1 million, with approximately 2 million visitors to the region during the summer months (25% of whom are families with children). Location near markets is extremely important and is reflected in industry research figures on the percentage of the local population that can be expected to go to the Water Park. When typical penetration rate data obtained from the World Waterpark Association are applied, a table of attendance projection figures is derived (see Table 2.1).

Although based on industry research, annual attendance figures are inherently difficult to project, particularly given the major impact that weather has on water park attendance. These penetration rates are considered conservative. Nevertheless, over the longer term, the estimated attendance is considered accurate to ±20%.

Admission Price Calculation

The admission price structure (see Table 2.2) is based on price structures in similar water parks in other locations in Canada and is com-

Table 2.1: Projected Annual Attendance

Residents			
Market Area	***Population in Segment Area***	***Penetration Rate (%)***	***Estimated Attendance***
0–8 km	200,000	40	80,000
8–16 km	300,000	30	90,000
16–24 km	280,000	20	56,000
24–40 km	320,000	10	32,000
Tourists			
Summer	500,000 (family members)	10	50,000
Total			***308,000***

petitive with other recreational attractions in the Ottawa–Carleton region.

Gate Receipts

Based on these calculations, annual estimated gate receipts would be $5,728,800, or approximately $5,729,000 (308,000 guests per year × $18.60 average admission price per person).

Table 2.2: Admission Price

Price Category	***Age (years)***	***Percent of Attendance***	***Price ($)*** ***Full Day***	***Price ($)*** ***Half Day***	***Prorated Price* 70:30 (Full:Half)***
Infants	0–2	~0	Free	Free	0
Children	3–12	25	16.00	12.00	3.70
Adolescents	13–17	35	20.00	14.00	6.37
Adults	18–64	39	24.00	16.00	8.42
Senior	Over 65	1	12.00	8.00	0.11
				Average price per person	**$18.60**

*Prorated price is determined by multiplying percent of attendance by [(0.7 × full day price) + (0.3 × half day price)]

Site Capacity

During the 15-week operating period from late May to early September, Meteorology Canada climatologic information indicates the following conditions:

Excellent	35 days	65 days
Good	30 days	
Fair	10 days	
Poor	30 days	

Assuming that almost all guests come during the best 65 weather days, the park would require a minimum site capacity of:

$$308{,}000 \text{ guests} + 65 \text{ days} + 1.3 \text{ turnover rate}^{*} = 3{,}645 \text{ guests capacity}$$

This figure assumes a continuous 100% capacity during best weather days. This is unrealistic. A better initial planning figure would be a **site capacity of approximately 4500 guests** (80% capacity required to meet attendance projections). Industry research also indicates that attractions must be expanded or "refreshed" every two to three years to attract return guests and maintain attendance figures. Future expansion potential of up to 7000 maximum capacity should be planned.

Parking Revenue Calculations

Industry figures indicate that 60 to 75% (average 67.5%) of guests arrive by car, with a further 10 to 25% dropped off and the remainder traveling by public transit, by bicycle, or on foot. Although all modes of transportation need to be considered in planning access to the Water Park, suitable parking is a key criterion for success.

With the average car containing 3.6 riders (industry figure) and a proposed initial site capacity of 4500 guests, parking for approximately 1000 vehicles including buses and recreational vehicles (RVs) is required.

$$[4500 \times 67.5\% + 3.6 = 843.8]$$

Estimated revenues, based on $5 per car would be:

$$308{,}000 \times 67.5\% + 3.6 \times \$5.00 = \$288{,}750$$

or approximately **$289,000** per year.

---

*As the water park would be open from midmorning to early evening, most guests would not stay all day, allowing a turnover of guests on any given day. Industry estimates are that a turnover rate of 1.3 can be expected.

Concession Revenues

Concession revenues consist of food, beverages, and merchandise as well as miscellaneous equipment rentals. Based on an industry average of $5 per person, annual revenues would total:

308,000 × $5.00 = **$1,540,000**

Total Projected Annual Revenues

Gate receipts	$5,729,000
Parking	$289,000
Concessions	$1,540,000
Total	**$7,558,000**

Cost Analysis

Initial Project Costs

This includes all costs associated with the planning, design, construction, and setup of the Water Park. The concept is for a project management team to undertake this work as a turnkey project, with eventual operation of the Water Park passed to an operations team.

Preliminary estimated project costs for a 4500-capacity water park are:

Preliminary planning and studies	$200,000
Site construction	$8,000,000
Attractions and equipment	$2,000,000
Initial marketing, promotions, and advertising	$300,000
Project management fees	$1,500,000
Total project costs	**$12,000,000**

These figures are based on industry estimates for a typical water park of this size plus an estimated 15% of construction and attraction/equipment for project management. Costs can be limited by reducing the number and quality of the attractions.

Future Refreshing of Attractions

Industry guidelines recommend that attractions be expanded or "refreshed" every two to three years to attract return guests and maintain attendance figures. It is estimated that **Can$1 million every three years** should be planned for this purpose. This cost is a discretionary figure in any single year, but the longer-term viability of the Water Park is contingent on this reinvestment.

Annual Operating Costs

The annual operating costs shown in Table 2.3 are based predominantly on industry data obtained from the World Water Park Association database. As most of the costs are variable, they have the same degree of accuracy as their underlying revenues, or ±20%. The fixed costs are somewhat more accurate and are estimated to be in the range of ±10% to 15%.

Evaluation

Capital Budgeting Methodology

The Net Present Value (NPV) methodology is used to calculate the economic viability of this project. The financial reference consulted was *Canadian Financial Management*, 4th edition (Brigham, Kahl, Rentz, and Gapenski; Harcourt Brace and Company, 1994). Note that in applying NPV calculations, depreciation expenses are not considered cash outflows, although they are used to determine income tax costs. Financing costs such as interest expenses also are not included in the project's cash flows as financing costs are implicitly incorporated in the company's cost of capital.

Table 2.3: Annual Operating Costs

Land lease (fixed):	$50,000	Based on preliminary discussion with regional authorities
5% of revenues from $4M to $7M	$150,000	
10% of revenues over $7M	$56,000	
Advertising and promotion: 10% of revenues	$756,000	Industry standard
Concessions: food, drink, merchandise 50% contribution margin	$770,000	Competitive pricing
Salaries: 15% of revenues	$1,134,000	Industry standard
Repairs and maintenance:		
Fixed: 3% of realty assets and attractions ($10M)	$300,000	Industry standards
Variable: 4% of revenues	$302,000	
Utilities: 5% of revenues	$378,000	Industry standard
Insurance (fixed):	$300,000	Preliminary discussion with insurance company
Administrative overhead (all fixed):		
Management salaries	$300,000	
Business and realty taxes	$70,000	
General administration and office	$500,000	
Total Fixed	**$1,520,000**	30% of operating costs
Total Variable	**$3,546,000**	70% of operating costs
Grand Total	**$5,066,000**	

Cost of Capital

It is assumed that capital financing, both debt and equity, can be obtained at an average nominal rate of 9%. It also is assumed that the inflation rate will remain at 2.5% throughout the period and will affect all revenues and costs equally. The real cost of capital to the company, then, is 6.5% (9% − 2.5%).

Net Present Value Calculations

A detailed cash-flow analysis and NPV calculation are given in Appendix A. Utilizing the factors summarized below, the **NPV result is Can $6,020,000**, indicating an economically viable project with a very good rate of return.

- Gross revenues of Can$7,558,000 based on a projected attendance of 308,000
- An initial project cost of $12,000,000 to set up the Water Park, with an additional $1,000,000 invested in new attractions every three years
- Annual operating costs of $5,066,000 (approximately 70% of which are variable costs)
- An average real cost of capital of 6.5% (nominal interest rate of 9%)
- Capital asset depreciation of 10% annually based on a declining balance

Sensitivity Analysis

Attendance Break-even Point

Keeping all other factors constant, the break-even attendance was determined to be 240,000 visitors annually. At this point, revenues would total $5,910,000 and operating costs would be $4,228,000, resulting in a Net Present Value calculation of zero. The break-even attendance represents a 22% reduction from the baseline calculation of 308,000 visitors and is on the edge of the estimated accuracy of ±20% for this figure.

Break-even Point of Cost of Capital

Using a similar approach as above, the cost of capital was varied, with other factors remaining at the base level, to determine the break-even point for financing at which time the project would no longer be viable. The break-even point (NPV = 0) was determined to be a real rate of 11.5% (nominal rate of 14%). Looked at another way, as long as the nominal rate for capital is less than 14%, the project remains viable.

Risk Assessment

In any project of this type, there are inherent risks. For the Water Park project, these risks can be divided into two segments: those associated with the planning, construction, and set-to-work of the Water Park and those associated with the ongoing viability of the park.

Planning, Construction, and Set-to-Work Phase

Financing

Although the NPV calculations indicate a very viable project with a potential for a strong rate of return (baseline calculations indicate 14%), uncertainty about investor financing remains a major risk area. The project has a significant capital exposure (approximately $12,000,000 to establish the water park) with no opportunity for revenues until the park goes into operation in late May 2005. The competition for investment financing remains high. In the Ottawa–Carleton region, with its strong high-technology base, most entrepreneurial funding is focused in that direction. A strong business plan and marketing effort will be required to obtain the financing needed to see this project through to the operational phase.

Construction Delays

The construction and set-to-work schedule needs to be set up to ensure that the park opens on the long weekend in May to take advantage of the entire summer season. Failure to open on schedule could result in cash-flow problems in that the project would be carrying financing costs without a source of income for an additional period. Further, the revenues in a shortened first summer of operation may be insufficient to cover costs until the park opens the following May. Good project management will be required to set up a workable schedule and then to follow through on that schedule.

Construction Cost Overruns

The preliminary project costs have been based primarily on standardized data obtained from the World Waterpark Association as well as from visits to other water parks in Canada. Once the location for the Ottawa–Carleton Water Park has been confirmed, tailored costs specific to the location will be required. Again, strong project management will be required to ensure that the project remains on budget. Only highly reputable firms should be contracted. Fixed-price contracting should be used to the maximum extent possible. Penalties should be written into the contract and enforced.

Operating Phase

Annual Attendance

The total number of customers annually is a key element in forecasting the financial viability of the Water Park. The projected annual attendance of 308,000 is based on a conservative interpretation of the best industry figures available. Furthermore, sensitivity analysis indicates that the break-even attendance of approximately 240,000 (22% less than projected) is highly achievable. A concerted promotional and advertising campaign needs to be undertaken to ensure that the park gets off to a good start when it opens. In addition, a major effort has to be made to maximize revenues through a combination of gate receipts and on-site revenues such as food and beverage concessions, merchandise sales, and various equipment rentals.

Weather

A detailed analysis of Ottawa–Carleton regional weather indicates that the Water Park will be very successful during a summer with average weather and can make a small profit in years when the weather is poor. A significant factor in this regard is that the Water Park is able to reduce daily operating expenses on inclement-weather days significantly by reducing staff for those days: Note that approximately 70% of operating costs are variable. Only the minimum of staff required to handle the number of customers would be maintained on-site. Seasonal water park staff would be paid only for hours actually worked.

Conclusion

Initial projections indicate that the Ottawa–Carleton Water Park would be a very viable project with a solid financial return. Estimates point to an annual attendance of 308,000 visitors generating revenue of approximately Can$7,600,000 and profits of $2,500,000. Assuming an average financing cost of 6.5% over inflation results in a positive NPV of $6,000,000 and an anticipated real rate of return of 11.5% (nominal rate of 14% assuming inflation of 2.5%).

There is no undue risk in this project that cannot be mitigated. The most significant risk is assessed as the ability to obtain the necessary financing given the highly competitive market for entrepreneurial financing. If this risk were to occur and the project were delayed or ultimately canceled, there would be relatively little financial exposure because the project would still be at the preliminary planning stage. Good project management will be required to keep the planning and construction phases of the project on schedule and within budget. Once secure financing is in place, overall project risk is reduced substantially. Risk during the operation phase centers on achieving attendance figures, a major component of which is the weather.

Adherence to the schedule is critical in this project. To ensure sufficient cash flow in the first year of operation, it is important that the Water Park commence operation in late May to capture all of the water park season. To avoid jeopardizing a May 2005 opening, the project will need to commence in early fall 2002 to provide the necessary time to obtain site approvals, capital financing, and design and complete the construction of the Water Park.

It is recommended to proceed with this project.

Appendix A: Water Park Cash-Flow Analysis

The cash-flow analysis is shown in Table 2.4

Table 2.4: Water Park Cash-Flow Analysis, Baseline ($000)

	Year 1	*Year 2*	*Year 3*	*Year 4*	*Year 5*	
Revenues (R)	7,558	7,558	7,558	7,558	7,558	
Operating Costs (OC)	5,066	5,066	5,066	5,066	5,066	
Depreciation (D)	600	1,140	1,026	1,023	921	
Taxable Income	1,892	1,352	1,466	1,469	1,571	
Taxes (@40%) (T = 0.4)	757	541	586	587	628	
Net Operating Income (NI)	1,135	811	880	881	943	
Add Back Depreciation (D)	600	1,140	1,026	1,023	921	
Reinvestment			(1,000)			
Project Net Cash Flow (CF)	1,735	1,951	906	1,905	1,864	
Net Present Value (at Year 0)	1,629	1,720	750	1,480	1,360	6,940

	Year 6	*Year 7*	*Year 8*	*Year 9*	*Year 10*
Revenues (R)	7,558	7,558	7,558	7,558	7,558
Operating Costs (OC)	5,066	5,066	5,066	5,066	5,066
Depreciation (D)	829	846	761	685	717
Taxable Income	1,663	1,646	1,731	1,807	1,775
Taxes (@40%) (T = 0.4)	665	658	692	723	710
Net Operating Income (NI)	998	988	1,038	1,084	1,065
Add Back Depreciation (D)	829	846	761	685	717
Reinvestment	(1,000)			(1,000)	

(Continued)

Table 2.4: Water Park Cash-Flow Analysis, Baseline ($000) (Continued)

	Year 6	*Year 7*	*Year 8*	*Year 9*	*Year 10*	
Project Net Cash Flow (CF)	827	1,834	1,800	769	1,782	
Net Present Value (at Year 0)	567	1,180	1,087	436	949	4,220

	Year 11	*Year 12*	*Year 13*	*Year 14*	*Year 15*	
Revenues (R)	7,558	7,558	7,558	7,558	7,558	
Operating Costs (OC)	5,066	5,066	5,066	5,066	5,066	
Depreciation (D)	645	581	623	560	504	
Taxable Income	1,847	1,911	1,869	1,932	1,988	
Taxes (@40%) (T = 0.4)	739	765	748	773	795	
Net Operating Income (NI)	1,108	1,147	1,122	1,159	1,193	
Add Back Depreciation (D)	645	581	623	560	504	
Reinvestment		(1,000)			(1,000)	
Project Net Cash Flow (CF)	1,753	727	1,744	1,719	697	
Net Present Value (at Year 0)	877	342	769	712	271	2,971

	Year 16	*Year 17*	*Year 18*	*Year 19*	*Year 20*	
Revenues (R)	7,558	7,558	7,558	7,558	7,558	
Operating Costs (OC)	5,066	5,066	5,066	5,066	5,066	
Depreciation (D)	554	498	449	504	453	
Taxable Income	1,938	1,994	2,043	1,988	2,039	
Taxes (@40%) (T = 0.4)	775	797	817	795	815	
Net Operating Income (NI)	1,163	1,196	1,226	1,193	1,223	
Add Back Depreciation (D)	554	498	449	504	453	
Reinvestment			(1,000)			
Project Net Cash Flow (CF)	1,717	1,695	675	1,697	1,677	
Net Present Value (at Year 0)	627	581	217	513	476	2,413

(Continued)

Table 2.4: Water Park Cash Flow Analysis, Baseline ($000) (Continued)

	Year 21	*Year 22*	*Year 23*	*Year 24*	*Year 25*	
Revenues (R)	7,558	7,558	7,558	7,558	7,558	
Operating Costs (OC)	5,066	5,066	5,066	5,066	5,066	
Depreciation (D)	408	467	421	378	441	
Taxable Income	2,084	2,025	2,071	2,114	2,051	
Taxes (@40%) (T = 0.4)	834	810	829	845	821	
Net Operating Income (NI)	1,250	1,215	1,243	1,268	1,231	
Add Back Depreciation (D)	408	467	421	378	441	
Reinvestment	(1,000)			(1,000)		
Project Net Cash Flow (CF)	658	1,682	1,663	647	1,671	
Net Present Value (at Year 0)	175	421	391	143	346	1,476
					Initial Investment Year 0	(12,000)
					NPV	**6,020**

Notes:

1. Depreciation is based on 10% annually on a declining-balance basis and a half rate in the year of acquisition. Note that three-year recapitalization causes variance in year-to-year depreciation.
2. A nominal tax rate of 40% is assumed.
3. NPV is calculated by discounting Project Net Cash Flow (CF) back-to-year 0 at a rate of 6.5% compounded annually. The general NPV discount formula is: NPV = Value in future year t\(1 + [Interest rate/100])^t where t is the number of years in the future.

3

Project Charter

Project Charter: Discussion

A project charter is the basis of a contract for an internal project; it is similar to a proposal (discussed in Chapter 4), which is a contract for an external project. A project charter is a planning document that is used to describe the project to managers and other concerned parties within the organization. It is similar to a project proposal in that it contains an initial step-by-step explanation of how the project will be built. Just as a project proposal is used to get a commitment from external clients to buy the project, a project charter is used to get a commitment from internal managers and other stakeholders to support the project. It provides the framework of an agreement between the project sponsors requesting the product or service and the project team that provides the deliverable. It defines the scope, objectives, approach, and deliverables at the outset of the project. A project charter, like a project proposal and a detailed project plan (see Chapter 5), is the responsibility of the Project Manager. Both the charter and the proposal are used to gain project approval from the sponsor agency, and both provide the basis of the detailed Project Plan that defines all project management requirements. Many aspects of the Project Charter will be detailed further in the final Project Plan; however, it is advisable to include as much information as possible in the Charter to assure the sponsor that the project will be successful.

Project Charter: Outline

Executive Summary

This is an optional section that does two things: It summarizes what is a project charter is, and it summarizes the key conclusions of the document. This gives the reader a chance to decide whether to read the document.

Part 1: Project Governance

Background

Provide the readers with a little history because they may not know anything about the project. State here how the project started, why, the original high-level objectives, the ballpark cost and the schedule. Some or all of this information may be contained in the Project Concept Document (see Chapter 1). Outstanding issues from the business case or project approval process to date need to be documented at the outset. Reference documents such as the Business Case (see Chapter 2) should be listed.

Project Purpose

This part provides a brief description of the project and its expected outcome or deliverable, identifying the final user of the deliverable and other stakeholders that will be affected. The Business Case (see Chapter 2) will probably have been completed by this time and may be an excellent reference as it provides a justification for the project's cost and benefit.

Project Objectives

The objectives that were stated broadly in the previous sections must be defined in detail, including all major products, services, and process deliverables. High-level functionality and performance may be mentioned here, as well as major constraints to be overcome. Ballpark cost and time and delivery dates normally are provided. Again, use the Business Case that was employed to sell the project as a reference. Objectives should be clear, tangible, specific, simple, realistic, and attainable.

Project Scope

In this section you should clarify both the project scope and the product or service scope. The project scope defines the work required to complete the product or service deliverable defined in the project purpose, and explains *how* this will be accomplished. The product scope outlines the capabilities and limitations of the product or service in general terms and should refer to the details in a related requirements or specifications document. You should include a high-level Work Breakdown Structure (WBS) to describe the work involved. There is normally a change/scope control strategy that allows the Project Manager to address deviations from the original project plan. Explain briefly the steps of change control: input, analysis, implementation, tracking, and configuration management.

Project Management Strategy

Your project management philosophy and approach should be outlined here. Issues such as contracting out some of the work, in-house capabilities, management style, project management methodology, quality, and professional standards can be addressed. Shortcuts or other imaginative approaches to achieve project objectives can be described. Project management software or other tools you are planning to use to control the project need to be described. The process and responsibility for project approval should be defined or recommended, depending on the composition of the project team.

The Project Team

List the names of the project manager and individual project team members—these people normally are identified as resources in the original business plan and/or project proposal—along with their general areas of responsibility. If the actual names are not known yet, state the knowledge level required for each position. This is important because there can be as much as an eight-to-one productivity difference between the most experienced and least experienced staff.

Diagram the organization of the Project Team (see Figure 3.1). Show the authority relationships (who reports to whom and the level of supervisory control—dotted-line versus solid-line authority). Typical roles include the project manager, project leader, and project team members in functions such as finance, design, legal, quality, and technical specialists. The organization's executive committee or senior management interface also may be included.

Detail the responsibilities of each member or job type and include a forecast of effort for each member of the team. Project Manager roles involve leading, motivating, and assigning responsibilities and tasks. Project managers are responsible for all outside communication (reporting, meetings, user or upper-level management interface) and ultimately are responsible for successful project completion. Team leaders supervise team members on technical details, manage the delivery of most complex activities, and control the technical quality of the product. Team members are responsible for their specific tasks and for reporting progress.

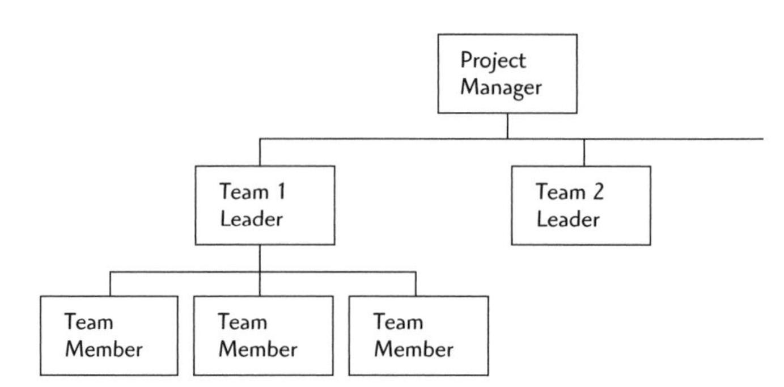

Figure 3.1 The Project Team.

Stakeholders

Client relationships. List all clients and their responsibilities, such as obtaining legal permission or approval and furnishing equipment. Also detail all stakeholders, their relationship to the project, their responsibilities, and their interfaces with your team. It may be useful to outline the risks involved if their roles are not fulfilled. This issue is developed further in the Risk Management Plan (see Chapter 7).

Outside dependencies. You need to identify clearly interfaces with internal or external agencies that are outside the control of the project manager and that may influence the project. Pay special attention to external (third-party) or internal agencies not directly concerned with the success of the project. List the stakeholders, their responsibility, the date you require the item, and the impact of not receiving it on time. All these interfaces also should be addressed in the Risk Management Plan (see Chapter 7).

Part 2: Project Details

Project Schedule

The project schedule should summarize the time required for each phase. Include a description of how the project and/or product milestones fit into the overall schedule. Provide a summary WBS listing the major project activities and duration—recall that time is a function of the task—and the resources available, including the size and capability of the team. Explain the major task dependencies and how the schedule was developed (such as backward from a deadline). Define the methodology for schedule control; take into account external dependencies that may be outside the project manager's control. Identify the project management software or manual tools being used to track the schedule.

List the major milestones in the schedule and the involvement of specific stakeholders. Include interim reports and deliverables. One important series of milestones is related to the bail-out points: interim project reviews or key actions (such as a loss of funding) that could lead the project manager or project sponsor to cancel the project.

Project Cost

Provide a high-level summary of project costs, most likely based on the business plan. Include meaningful cost roll-ups: by phase, deliverables, or locations. Explain the accuracy or range of values and show how these estimates were calculated. Define the concept for cost control, including signing authority and levels of approval for the project team and the project sponsor.

Quality Control

Explain the process for defining and achieving quality standards. Outline the strategy for management and technical reviews, inspection, testing, and other quality control and assurance activities. Include quality control of both project deliverables and internal project management processes.

Risk Management

Complete a summary of risks and include an overview of the Risk Management Plan (see Chapter 7). Explain the strategy for controlling risk and mitigating the consequences of potential problem areas, along with contingency plans for these risks. Key areas of concern may include finances, project resources, schedule delays, and quality control issues.

Communications Plan

Describe the broad communications objectives, key messages, and audiences. List those who will receive information, when, why, and how. Describe the method of communication. Describe the procedures and formats for internal (project team) and external (stakeholders) communications. Project communications ensure that all concerned parties are updated on project status and on changes to the project's scope, cost, or schedule that occur during the development of the project.

Support Activities

Describe any requirements for project support activities that will assist in the transition of the project from development to ongoing operations. These activities include user training, long-term quality assurance, and operational manuals and procedural documents.

Documentation

Explain the two classes of documents in the project: user and project management documents. Outline what documents will be produced, when, and the responsibilities for production and approval.

You will have to be very specific about the reporting format and content of the status reports, milestone reports, and other project documents.

Detail who will create the reports, the frequency, who will receive each report, when, and what the receiver's responsibility is after receiving the report. Again, be conscious of the impact on the schedule of outside agencies that either prepare or must accept these reports.

Project Administration

Detail the facilities and resources required from the project team, including office space, communications equipment, and computer hardware and software. Include an outline or reference for administrative procedures to be followed by the project team.

Terms and Conditions

Describe the legal, ownership, liability, and contractual terms and conditions of the project. These references often are provided by the organization's legal and finance staff.

Assumptions

List any assumptions and potential risks if those assumptions turn out to be false.

Summary

Repeat the three or four most important messages in the Project Charter, normally related to the project aim and objectives, the outline project plan, and the resources involved.

References

Provide details of key reference documents, including approval and decision documents, relevant publications or communications, and theoretical or academic material.

Appendixes

List appendixes of details or amplifying material, including the project WBS, financial data, risk issues, and other background detail. Provide a list of acronyms and abbreviations with explanations if that is required.

Project Charter: Example

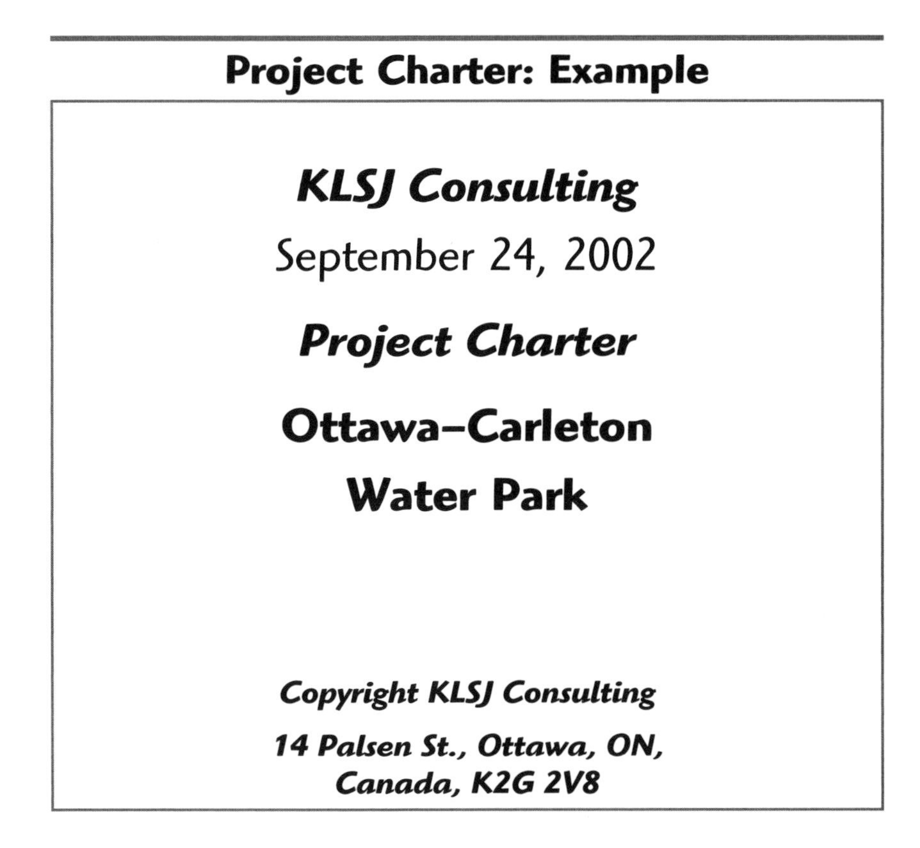

KLSJ Consulting

September 24, 2002

Project Charter

Ottawa–Carleton

Water Park

Copyright KLSJ Consulting

14 Palsen St., Ottawa, ON,
Canada, K2G 2V8

Executive Summary

The following document provides a *Project Charter* for the Ottawa–Carleton Water Park to be developed by KLSJ Consulting. The charter describes the background and objectives for the project and explains the strategy behind the project management structure proposed by KLSJ Consulting. To meet the proposed targets, *approval of this Project Charter by the owners of Carlington Aquatic Parks is required* within 30 calendar days *from the date of this report.*

The *management philosophy* for the project is as follows: KLSJ Consulting will manage the design, development, and construction of the park. KLSJ will hand over operations to the operations management team before the official opening. Carlington Aquatic Parks will retain approval authority for all critical design, marketing, and financial decisions throughout the project. Project *team resources* include the Project Manager, Project Leader (Design and Construction), Team Leader (Legal), and Team Leader (Financial) as well as a Risk Manager. The design and construction team also includes three junior members for the construction period. A contracted firm overseen by the Project Manager will carry out marketing.

The Work Breakdown Structure (WBS) shown in Appendix A describes the high-level activities and schedule for the project, with the top two levels showing phases and then functions within each phase. Subsequent planning activities will breakdown these categories further into tasks. In some cases project tasks may still be rather lengthy; therefore, a subordinate WBS with increased granularity will be developed for those areas. The schedule also identifies the *critical path* of activities and milestones that must be met to achieve the target opening date of May 21, 2005.

Project duration is from September 2002 to May 2005, a period of 32 months. The total project cost is Can$12,450,000, based on a Class B estimate, with a range of +25% to −10%.* The project risk is high for Phases 1 (Concept) and 2 (Planning), but once the issues of land approval and capital investment have been addressed, the project risk becomes low for Phase 3 (Execution) and beyond. Risk management and mitigation are dealt with at length in the document, along with alternatives for balancing the requirements of time, cost, and quality.

Contents

---

*Max Wideman, http://www.maxwideman.com/issacons3/iac1332/tsld004.htm, January 15, 2002. A Class B estimate is defined as having a value within a range of +20% to −10%.

Appendixes

Part 1: Project Governance

Background

Dan Milks, president and CEO of Carlington Aquatic Parks of Ottawa, has identified the *market need and potential benefits* of a water park in Ottawa, Canada, also known as the National Capital Region. Mr. Milks envisions a family-oriented water and amusement facility that would provide a recreational outlet for visitors and local residents. Ottawa currently lacks this type of outdoor recreational attraction. Mr. Milks issued a *Project Concept* document in September 2002 to market his idea to the local government and to investors. The plan is to open by spring 2005. Cost is about Can$12 million with a further $4 million earmarked for future development.

Given the substantial support throughout the region for the idea, Carlington Aquatic Parks has decided to proceed with the development of the Water Park project. KLSJ Consulting has developed the *Project Charter* to map out the process for the successful and timely completion of the Ottawa–Carleton Water Park.

Purpose

The purpose of this Project Charter is to describe how KLSJ Consulting will complete the design and construction of the Ottawa–Carleton Water Park facility on behalf of Carlington Aquatic Parks. The Water Park will be completed and ready for opening May 21, 2005, at a total cost of $12,450,000. Note that this cost is a Class B estimate with a range of +25% to −10%. To meet the proposed targets, approval of this Project Charter by the owners of Carlington Aquatic Parks is required within 30 calendar days from the date of this report.

Objectives

The objective of the project is for Carlington Aquatic Parks is to create a top-quality recreational facility with the following features:

- Capacity of 4500 guests per day at opening, with growth potential for up to 7000 guests per day
- Unique attractions, including an outdoor wave pool, an artificial river, and water slides

- A proposed site in the western portion of Ottawa–Carleton's greenbelt, within a 45-minute drive for over 1 million residents of eastern Ontario and western Quebec
- A parking area adjacent to the Water Park capable of accommodating 1000 cars
- Serviced by local roads, bus transportation links, and bike trails

Project management theory speaks of a triple constraint: time, cost, and quality. Considering the fixed target date for the opening of the park and the low probability that extra funding could be located on short notice, any project-related problems probably will result in a decrease in the number of attractions or their quality. KLSJ Consulting can provide and execute an effective project management system that will ensure that the intended cost, time, and quality are delivered.

The major deliverables from the project include the following:

- Feasibility Study
- Business Plan
- Project Plan and Project Budget
- Risk Management Plan
- Environmental Assessment
- Zoning Approval
- Architectural and Engineering Design
- Project Presentations (for political and community groups)
- Investor Commitments
- Marketing Plan
- Operations Staff Guidance (financial, marketing, operations)
- Project Closing Final Report

Scope of Activities

The *WBS* for the Ottawa–Carleton Water Park is contained in an MS Project file and is shown in Appendix A. The plan contains five phases and covers a period of nearly 32 months. When the final plan is completed, the activities will be subdivided into three levels: phase, function, and specific tasks. Grouping activities by phase allows the Project Manger to get an overview of the interrelationships of different activities as well as the overall flow of the schedule. Grouping activities by function allows the Project Manager and other team members to track the interaction and sequencing of activities in a specific functional area. The project is initially subdivided into five *phases*:

- *Phase 1: Concept.* This phase involves developing and assessing the feasibility of the overall project concept. Key activities include the owners' partnership agreement and the initial feasibility study.

- *Phase 2: Planning.* This phase involves the development of plans and strategies required in managing the project. Key activities include creating the project plan and business plan, completing preconstruction (engineering) studies, and locating investors.
- *Phase 3: Execution.* The third phase involves the actual construction of the facility and also includes an increased focus on marketing.
- *Phase 4: Handover to Operations.* This phase involves the process of finalizing the facility, preparing it to open, and turning over operational control to the park manager.
- *Phase 5: Closing.* This phase involves internal project management functions required to close the project, including the project review and final client report.

The project also was subdivided into six *functional areas*, each the responsibility of a project team member, as shown in Table 3.1.

To keep the WBS activity list under 200 items (as specified by the project sponsor), many of the individual tasks are of much longer duration than the typical three to five days. Therefore, the *Master WBS* will be augmented by a series of six subordinate WBSs, one for each of the six major functional areas, to define the specific activities involved in each of these tasks more clearly.

A formal scope control process will be put in place to manage changes in scope as part of the overall change control system. Change cutoff dates are scheduled in each phase to allow changes to be incorporated early enough to minimize disruption to other WBS activities. A formal review and report to the client at the end of each phase provides a second opportunity to review changes to the scope of the project.

Table 3.1: Functional Areas and Project Team Responsibility

Function	***Responsibility***	***Majority of Effort***
Project management	Project Manager	Phases 1 to 5
Contract management	Team Leader (Legal)	Phases 2 and 3
Financing	Team Leader (Finance)	Phases 2 and 3
Political/legal	Team Leader (Legal)	Phases 1 and 2
Construction	Project Leader (Design and Construction)	Phases 2 and 3
Marketing	Project Manager (outside contract)	Phases 3 and 4

A scope management plan integrated with other control processes (schedule, cost, and quality) will be promulgated. This plan will have the following general characteristics:

- Change requests must be written and fully substantiated.
- Each change request must be assessed as to impact on cost and schedule.
- A tracking system will be implemented to monitor changes.
- Approval levels necessary for authorizing changes will be established.

Project Management Strategy

The *overall project strategy* calls for the initial planning and preparation to be completed by the end of Phase 2, before construction begins in Phase 3. Specifically, land approval, investor commitments, and conceptual design will all be completed before formal construction is undertaken.

The intention of Carlington Aquatic Parks is that KLSJ will assume all *responsibility* for coordinating the management of the development and construction activities. Carlington will remain actively involved in marketing the project to political offices, community groups, potential investors, and future customers. The Carlington Executive Committee will *retain approval* authority for all strategic aspects of the project unless they are specifically delegated to KLSJ. Thus, Carlington is expected to approve financial plans and major expenditures (above Can$10,000), investors, architectural and engineering designs, and marketing activities.

KLSJ will act as overall *project manager* while contracting specific aspects of project execution to technical firms selected by KLSJ and approved by Carlington. Functions to be contracted to outside experts include the following:

- Environmental assessment (including environmental, architectural, and traffic studies)
- Engineering site services study
- Engineering design and construction
- Marketing

Also, KLSJ will locate candidate firms for ongoing operations management of the facility.

KLSJ works exclusively with *Microsoft* software, including *Project* for project management, *Excel* for financial reports, *Word* for reports, and *PowerPoint* for presentations. All contractors will be expected to integrate with these systems. The specific information technology (IT) operating systems for ongoing facilities management will be determined during the development of the project.

Project Team

The *Project Team* will consist of five core members for the duration of the project, plus three additional persons required to coordinate activities during construction (Phase 3, Execution). The basic organization of the team is shown in Figure 3.2. Close horizontal communications among the team members will be essential.

The specific *responsibilities* of the team members are as follows:

- *Project Manager (Karen Dhanraj)*, General Manager of KLSJ. Will have overall responsibility for all aspects of the project on behalf of the CEO and ownership group of Carlington Aquatic Parks. The Project Manager has primary responsibility for the following areas:
    - Coordination and communication with outside agencies
    - Interface with community groups, the National Capital Commission (NCC), and municipal, regional, and provincial governments
    - Providing direction and guidance to other team members
    - Maintaining the overall project plan (schedule, cost, and resources)
    - Coordinating and producing all project documentation
    - Coordination and oversight of the marketing agency (outsourced)
    - Membership in the Risk Management Working Group
- *Team Leader (Design and Construction) (Scott Kennedy)*. Responsible to the Project Manager for coordination of all activities related to design, engineering studies, permits, and construction. The major goal of the Project Leader is to produce a high-quality water park. Specific responsibilities include the following:
    - Supervision of the preliminary design and scale model as well as detailed final design

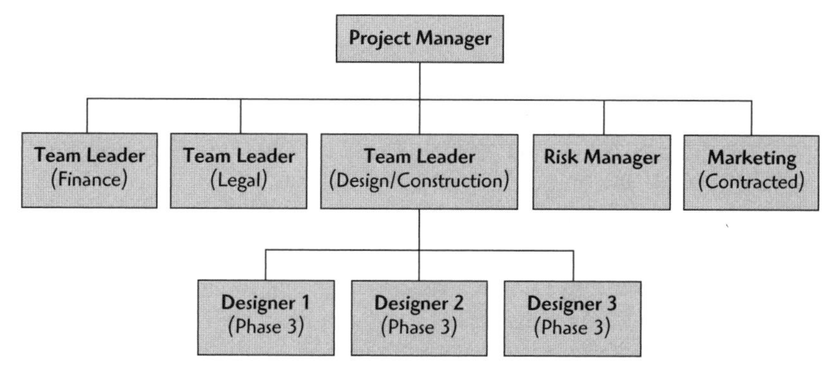

Figure 3.2 Organization of the Project Team.

- Coordination of the physical construction of the Water Park and related facilities
- Oversight of the various contractors involved in the construction phase
- Overall quality control of the Water Park
- Change control with respect to design and construction
- Supervision of the three additional project team members during Phase 3 (Execution)
- Membership in the Risk Management Working Group

▲ *Team Leader (Legal) (Laverne Fleck)*. Responsible to the Project Manager for legal counsel related to project activities as well as contract management and interface with regulatory authorities on legal issues. Specific responsibilities include the following:

- Drafting, review, and management of all contracts with contractors (contract management)
- Negotiation of long-term site lease
- Membership in the Risk Management Working Group

▲ *Team Leader (Finance) (Steve Jackson)*. Responsible to the Project Manager for financial issues related to the project. Specific responsibilities include the following:

- Development and monitoring of business plans, budgets, and project cash flows
- Lead responsibility for locating and confirming investors in conjunction with the president of Carlington Aquatic Parks, the ownership group, and banks and investment firms
- Payment of all project expenses incurred by project staff
- All project information technology (IT) systems both for internal project use and for ongoing use by the Water Park operations
- Membership in the Risk Management Working Group

▲ *Risk Manager (Jim Harris)*. Responsible to the Project Manager for coordinating the Risk Management Program. Specific responsibilities include the following:

- Advising on risk management issues
- Ensuring risk management documentation is current
- Chairing the Risk Management Working Group

▲ *Team Members (Design and Construction)*. Responsible to the Team Leader (Design and Construction) for overseeing specific aspects of the Water Park construction program. Workload analysis indicates that three persons will be required, each responsible for one of the following areas:

- Procurement, installation, and related construction of attractions (slides, pools, etc.)
- Construction and renovation of on-site facilities (buildings, lighting, landscaping)
- Construction of off-site facilities (i.e., parking lot, connecting roads)

Stakeholders

The sheer size and cost of the project imply that a large number of organizations and individuals will be directly involved or indirectly interested in the development of the park. A list of key *stakeholders* includes the following:

- Owners
- Investors
- Government (provincial, regional, and municipal, including the City of Nepean, the National Capital Commission, and the regional, provincial, and municipal governments)
- Community groups
- Contractors
- Employees
- Related business partners
- Competitors of the Water Park

Part 2: Project Management

Project Schedule

The project is scheduled to start on September 16, 2002, with the grand opening of the Water Park expected on Saturday, May 21, 2005 (Victoria Day weekend). Initially, the schedule was developed forward from a set start date. This resulted in an opening date in February 2005 with an unrealistic winter construction period. The schedule then was worked backward from the 2005 Victoria Day weekend to arrive at a workable project. For a more detailed depiction of the schedule, see the initial Work Breakdown Structure in Appendix A.

The WBS is divided into five phases. Phase 1 mainly involves determining the feasibility of the project. Once the viability of the Water Park has been determined, the project can proceed to Phase 2, in which zoning approval and investor capital are sought. With these attained, Phase 3, consisting of construction activities, can commence. Successful completion of construction permits the handover of the Water Park to the operations team in Phase 4. Once operations are in place, KLSJ can close the project in Phase 5.

Key milestones from the critical path within each of the five phases are shown in Table 3.2.

In addition, there are *other important milestones* that are not part of the critical path yet are essential for schedule progression, which are shown in Table 3.3.

Table 3.2: Key Milestones on the Critical Path

Milestone	*Date*	*Key Stakeholder*	*Key Trigger Event(s)*
Phase 1: Concept			
Approval of project team assignments	September 16, 2002	Project Manager	Creation of core project team
Selection of preliminary Water Park site	January 21, 2003	Project Manager	Assessment of site options
Approval of Phase 1 and end of phase review	March 17, 2003	Project Manager	Presentation to NCC authorities
Phase 2: Planning			
Approval of project charter by Carlington Executive Committee	April 28, 2003	Project Manager	Approval of Phase 1 (Concept)
Approval of environmental assessment	February 2, 2004	Project Manager	Completion of environmental study
Zoning approval from City of Nepean	February 2, 2004	Project Manager	Conduct environmental assessment, NCC site approval
Approval of Phase 2 and end of phase review	February 2, 2004	Project Manager	Zoning approval, confirmation of investors

Phase 3: Execution			
Approval of construction permit	TBC*	Project Leader	Completion of detailed engineering design
Approval of construction	TBC	Project Leader	Completion of construction, installation of attractions, landscaping
Approval of Phase 3 and end of phase review	April 19, 2005	Project Manager	Approval of construction
Phase 4: Handover to Operations			
Trial opening date	TBC	Project Manager	Final site cleanup and set-to-work preparations
Handover to the operating team	TBC	Project Manager	Posttrial modifications
Grand opening ceremony	May 21, 2005	Operations Team	Handover to operating team
Phase 5: Closing			
Completed postproject review	TBC	Project Manager	Handover to operating team
Project completion	June 7, 2005	Project Manager	Project presentation to client, financial reconciliation

*TBC: to be confirmed.

Table 3.3: Important Milestones *Not* on Critical Path

Milestone	*Date*	*Key Stakeholder*	*Key Trigger Event(s)*
Phase 1: Concept			
Confirming team member participation	November 11, 2002	Project Leader	Create core project team
Approval of the architectural design	June 23, 2003	Project Leader	Approve Phase 1 (Concept)
Phase 2: Planning			
Investors' commitment	December 8, 2003	Team Leader (Finance)	Approve detailed business plan
Site approval from NCC	December 22, 2003	Project Manager	Presentation to NCC

Each phase concludes with an end-of-phase review and approval milestone. In addition to these built-in schedule controls, KLSJ will use the following:

- A schedule change management plan that controls changes to the schedule caused by external influences
- Performance reports before the end of each phase, providing information about which planned dates have been met and which have not
- Change requests, required before the deadline for changes, that will alert KLSJ to schedule delays or accelerations

Project Costs and Financial Analysis

Cost Summary

The project is estimated to cost **Can$12,448,250**, which is a **Class B estimate** (indicating a possible range of −10% to +25%). This amount includes a 10% profit margin for KLSJ, as agreed on with the customer, as well as applicable taxes. KLSJ has confidence in this estimate, because the majority of the figures used are based on catalogue prices for equipment and similar construction work undertaken at other water parks. However, some of the larger cost estimates (for example, excavation and landscaping) are based on "rule of thumb" calculations for the magnitude of work specified. Where applicable, a "contingency" of 10% has been built into the cost estimates to account for anticipated variation in prices.

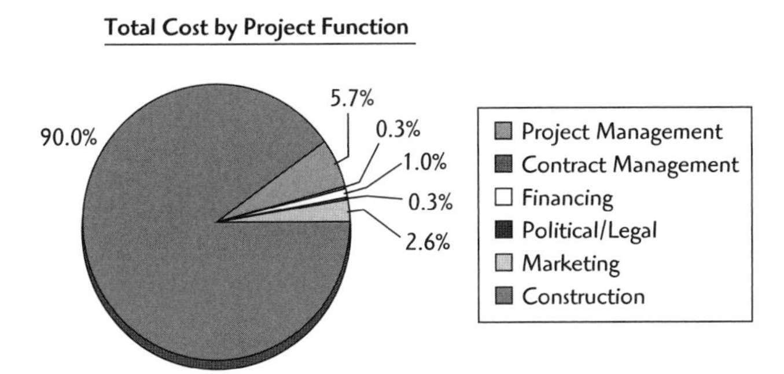

Figure 3.3 Total cost by project function.

Costs by Function and Phase

Figures 3.3 and 3.4 depict the costs by project function and phase. They highlight the fact that the vast majority of project costs will occur in Phase 3, during the actual physical construction of the Water Park, as one would expect. This also emphasizes the fact that the project *should not* proceed beyond Phase 2 until 85% of the funding has been secured. All project activity will cease at the end of Phase 2 until investors are formally committed, as discussed further under "Risk Management." Estimates of the project costs by project phase, project category, and cost category (chart of accounts) are provided in Appendix B. Two items of interest to the client are:

- Project management and contract management overhead account for up to 10% of the total cost, not including the profit margin.
- Only 4.5% (Can$511,550) of the total cost is expended during Phases 1 and 2, meaning that there is relatively little financial risk exposure until the project is committed to begin Phase 3.

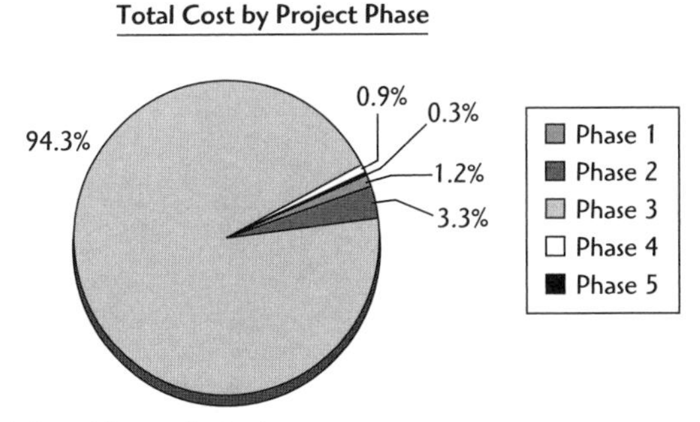

Figure 3.4 Total cost by project phase.

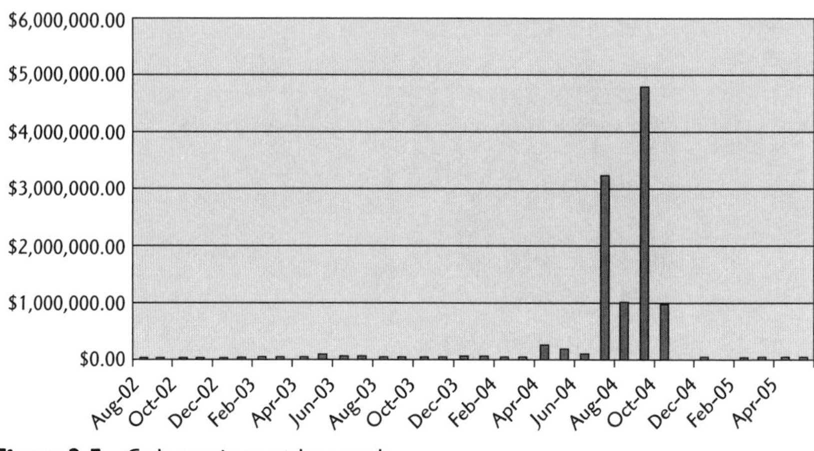

Figure 3.5 Cash requirement by month.

Cash Flow and Cash Management

Figures 3.5 and 3.6 depict the monthly and cumulative cash requirement for the project, based on current cost estimates and normal contracting payment policies. In cases where activities were more than one month in duration, it was assumed that roughly equal payments would be made at each month end based on specified contract performance. The figures again point out the imperative to secure all funding before the commencement of Phase 3. This funding will need to be liquid and readily available to KLSJ at all times to make contractor payments. Failure to meet the scheduled payments with contractors may result in work stoppage or even legal action against the client. For instance, during the four months between July and October 2004, the project will consume approximately Can$2 million per month, the majority of which will be for construction and materials. Details of access to bank accounts and other sources of funding will be the subject of separate discussions.

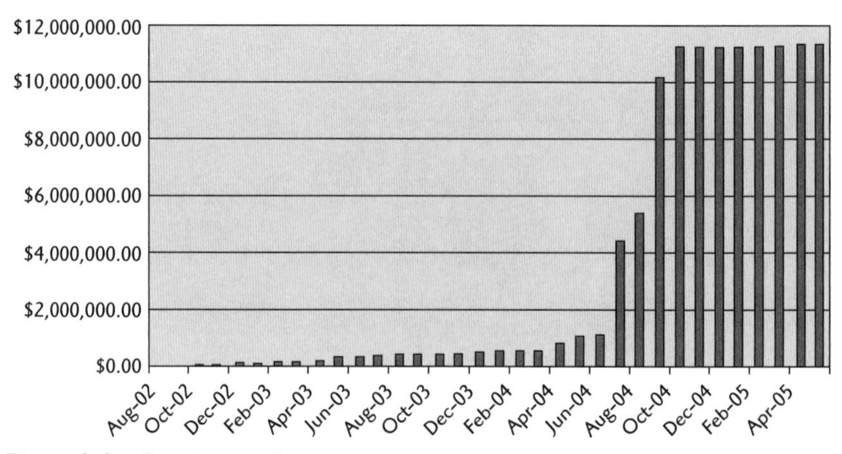

Figure 3.6 Cumulative cash requirement.

Cost Control Strategy

Other than the specific risk management strategies outlined separately, KLSJ will implement specific actions designed to minimize the possibility of cost overruns. These actions will include but are not limited to the following:

- Fixed-price contracts with contractors
- Substantial penalty clauses for nonperformance of any material aspect of a contract
- Where prudent, insurance coverage against specific perils (third-party liability, acts of God, nonperformance of high-risk contracts)
- When necessary and in consultation with the client, deliberate trade-offs between time, quality, and cost to remain within the budget.

Resource Requirements

The WBS list of activities was utilized to estimate the personnel resource requirements needed to manage the project. Each activity was assigned to the appropriate member(s) of the project team with an estimate of the level of effort required for each activity.

Project management resources have been costed at standard company charge-out rates, which include overhead and profit. These costs have been incorporated into the total project cost and are captured separately in cost account C011 (see Appendix B). Resource costs have been levied only for those periods in which a team member is engaged in project business. Finance and legal members, for example, are not required full-time on the project.

Quality Control and Performance Measures

The Project Team will employ a combination of qualitative and quantitative methods to monitor the project. Quality control measures ensure that the project is able to meet the specifications intended for the facility. The following quality control measures will be implemented to ensure that the project meets all customer needs and specifications. They will be performed by the project management team, the contractors, or indepen dent outside agencies as appropriate:

- *Control of risk* by the Risk Management Working Group
- *Control of suppliers, contractors, and subcontractors* through a "qualified seller" listing
- *Control of work performed* through inspection, procurement audits, destructive testing (concrete), and nondestructive testing (attractions)

- *Control of design and engineering* through strict document control, document change control, formal review processes, and independent consultation where necessary
- *Control of financial transactions* through segregation of responsibilities, expenditure controls and limits, limited payment authority, and an independent audit
- Where applicable, the *use of statistical tools* such as process capability and statistical process control to ensure that equipment and processes operate within specified tolerances

Risk Assessment

Recognizing and managing potential risks before they impact the project is an essential aspect of project management. Risk management is a proactive and iterative process that is vital to controlling costs, meeting deadlines, and producing quality results.

KLSJ Consulting has conducted an initial risk assessment of the Water Park project. The criteria for evaluating risk are presented in Appendix C, along with an initial list of the key risks that have been identified. Complete details of the risk assessment are outlined in the Risk Management Plan.

Based on the risk assessment and the high risk level in the early phases, KLSJ assesses the Ottawa–Carleton Water Park project as a high-risk project. The two most significant risks that the project faces are:

- Failure to secure funding in time to meet project deadlines
- Political action group(s) successfully petitioning the Zoning Authority, the Capital District, the Region, or the City of Nepean, resulting in delays or denial of site approval and subsequent delays in the start of Phase 3 (Execution) to the extent that completion by the target date becomes improbable

Each of these risks has a high probability of occurring and would have a large impact on the project both in terms of time delays and in terms of additional costs (see Appendix C, Table 3.5). Without commitment from investors, the project will come to a halt at the end of Phase 2.

Similarly, zoning approval from the City of Nepean is essential for the commencement of Phase 3. If zoning is not approved, not only will Carlington Aquatic Parks need to find a new location but the preliminary water park design and scale model may have to be redone completely. A change in location may result in withdrawal of potential investors and further battles with political groups that oppose an alternative site. In addition, the project team would need to reassess the financial requirements relative to the new site, which may render the current cost estimate invalid.

In an effort to minimize the impact of these two high-risk items, the project has been planned so that only a minimal financial commitment is required until these hurdles are overcome in Phase 2. The estimated cost to get to this point of the project is Can$513,550 (4.5% of project cost).

Based on the analysis above, the initial expenditure of approximately $515,000 is a high-risk investment. However, once financing has been secured and political approval has been received for the land and zoning, the overall level of risk for the Water Park drops significantly (see Appendix C, Table 3.6). At the start of Phase 3 (Execution), the remaining $11,500,000 is considered a low-risk investment.

A Risk Management Working Group (RMWG) has been established and will be responsible for all risk-monitoring and mitigation activities. The group will meet monthly at a minimum to review the risk forms and react appropriately. All changes will be published. Special meetings will be arranged to address emergency items as required.

Although the initial risk management process has identified many potential project risks, new risks will continue to surface during the two and a half years of the project. The Risk Management Team will monitor the probability and impact of risks constantly and revise the project completion date, costs, and deliverable dates accordingly. Carlington Aquatic Parks will be consulted and advised on such changes if and when they occur.

Communications Plan

The Communications Plan for the project addresses both internal communications among the project team members and external communications with other stakeholders. Internal communications include a routine weekly project team coordination session at which issues ranging from work progress and technical problems to administrative questions are addressed. This session is the primary means for the team to discuss and coordinate the impact of changes to project scope or schedules. More frequent planning and integration sessions can be conducted as required.

The purpose of external communications is to provide status reports to stakeholders and receive their feedback on an ongoing basis. Continuous stakeholder input can allow for properly controlled changes that prevent major problems downstream in the project. Communication methods should include a combination of static updates (reports and Web pages) along with interactive sessions such as project briefings. External communications should be scheduled around prominent milestones or at the completion of major project phases.

Support Plans

The KLSJ team will provide support plans to assist in the transition of management control from the project team to the operations staff. Along with the risk, communication, and quality plans included here, the Project Team will arrange for the development of user manuals for rides and training material for park staff. KLSJ also will ensure that all necessary facilities maintenance plans are provided by the contractors installing equipment in the park.

Related Project Plans and Documentation

The project staff and contractors will require several management documents during the project. All project documents will be made available to the client and the relevant operations staff at specified milestones or at the end of the project.

Other subordinate plans in support of the overall project plan include the following:

- Feasibility study
- Environmental assessment (including environmental, archaeological, and traffic studies)
- Engineering site services study
- Engineering design
- Business plan
- Financial management plan
- Risk management plan
- Marketing plan
- Operating plan (ongoing operations)
- Conceptual design (Appendix D)
- Complete functional work breakdown structure with the following sections:
  - Project management
  - Contract management
  - Political/legal
  - Financial
  - Construction
  - Marketing

Other project documents include:

- Partnership agreements
- Interim (end-phase) and final (end-project) client reports and presentations
- Procurement documents and contracts
- Warranties, clear title, deeds, licenses, registrations, and lease agreements

Project Administration

The project team will operate out of the regular KLSJ office facilities and will be allocated offices A-24 to A-31. The KLSJ administrative and accounting staff will provide administrative support to the project as required. On-site activities will be approved and funded by the Project Manager out of the project budget.

Terms and Conditions

The core project team will be operating as full-time members of KLSJ Consulting. Temporary project staff will be hired under the standard temporary employment conditions of KLSJ. KLSJ will enter into formal contracts with all suppliers and subcontractors, who will retain liability for all work and equipment provided to the project. Financing for the project will be the responsibility of Carlington Aquatic Parks VP of Finance, who will approve all contractual agreements with investors and suppliers. KLSJ will assume liability only for any loss or damage resulting from park design or development activities directly attributable to KLSJ or its employees.

Assumptions

The project plan described in this report is subject to a vast number of external and internal influences, many of which cannot be projected accurately or defined at the outset of the project. The project plan is founded on the following key assumptions:

- Economic and political environmental factors will continue to support the project concept.
- Community support for construction and ongoing patronage will continue.
- Sufficient debt and equity investors can be located no later than mid-February 2004.
- The budgeted amount of Can$12.45 million will be sufficient to complete the project as intended.
- The project start-up in September 2002 and opening in May 2005 are fixed milestones.
- A suitable location can be identified and acquired in a timely fashion.
- The environmental assessment will be favorable or mitigation can be made with reasonable effort.
- The investors will be satisfied with the engineering design.
- Contractors of sufficient size and expertise are available.
- Delivery of all equipment is within the time allotted.

- All permits and permissions are secured within the time allotted.
- The weather during Phases 3 and 4 remains within normal seasonal parameters.
- A suitable operations manager at a reasonable salary can be found.
- The trial run reveals no major problems.

KLSJ believes that these assumptions are reasonable. If any of the assumptions prove otherwise, however, there undoubtedly will be an impact on project cost, time, and/or quality. A risk analysis, as described previously, has been completed, and mitigation action as applicable will be undertaken to minimize these impacts.

Summary

KLSJ Consulting has developed a comprehensive Project Charter that will allow it to oversee the conceptual development, design, approval, construction, testing, and handover of the new Ottawa–Carleton Water Park. The Project Charter will maximize the probability of completing the facility as designed, on time, and on budget.

The first step on the road to the completion of this exciting new facility is the approval of the enclosed Project Charter by the Executive Committee of Carlington Aquatic Park. Approval is required within the next 30 days in order for KLSJ to maintain the project schedule.

References

1. "Few Waves Yet to Ride," *Link Magazine*, Spring/Summer 1999.
2. Rakos, J., *Software Project Management for Small to Medium Sized Projects*, Prentice-Hall, Englewood Cliffs, NJ, 1990.
3. Project Management Institute, *A Guide to the Project Management Body of Knowledge (PMBOK) Guide*, Newton Square, PA, 2000.
4. Wideman, Max, http://www.maxwideman.com/issacous3/iac1332/tsld004.htm, January 15, 2002.

Appendix A: Summary WBS and Schedule

This summary Work Breakdown Structure and schedule are shown in Figure 3.7.

ID	Task Name
1	OC Water Park
2	Phase 1 - Concept
3	Project Management
14	Contract Management
17	Financing
19	Political/Legal
22	Construction
25	Marketing
27	Phase 2 - Planning
28	Project Management
47	Contract Management
54	Financing
58	Political/Legal
68	Construction
71	Marketing
75	Phase 3 - Execution
76	Project Management
84	Contract Management
94	Financing
99	Construction
124	Marketing
131	Phase 4 - Turnover to Operations
132	Project Management
139	Contract Management
141	Construction
144	Phase 5 - Closing
145	Project Management
150	*Project Complete*
151	Financing

Figure 3.7 Summary WBS and schedule.

Appendix B: Financial Estimates

Financial estimates for the park are shown in Table 3.4.

Table 3.4: Ottawa–Carleton Water Park Costs (Can$)

Totals by Project Phase

	Fixed Costs	*Resource Costs*	*Total Costs*
Phase 1. Concept	$25,000	$109,000	$134,000
Phase 2. Planning	$192,000	$185,550	$377,550
Phase 3. Execution	$10,584,000	$134,100	$10,718,100
Phase 4. Handover to Operations	$70,000	$28,500	$98,500
Phase 5. Closing	$0	$33,000	$33,000
Total	**$10,871,000**	**$490,150**	**$11,361,150**
Profit (10%)	$1,087,100		$1,087,100
Grand Total	**$11,958,100**	**$490,150**	**$12,448,250**

Totals by Project Category

	Fixed Costs	*Resource Costs*	*Total Costs*
Project Management	$432,000	$213,050	$645,050
Contract Management	$0	$38,300	$38,300
Financing	$34,000	$83,200	$117,200
Political/Legal	$0	$35,400	$35,400
Construction	$10,135,000	$91,100	$10,226,100
Marketing	$270,000	$29,100	$299,100
Total	**$10,871,000**	**$490,150**	**$11,361,150**
Profit (10%)	$1,087,100		$1,087,100
Grand Total	**$11,958,100**	**$490,150**	**$12,448,250**

Totals by Cost Category

C011. Salaries	$490,150
C022. Site Visits	$5,000
C031. Marketing Services	$45,000
C032. Advertising	$200,000
C034. Promotions	$25,000
C045. Studies	$77,000
C054. Permits	$75,000
C061. Design and Engineering	$365,000
C062. Construction—On-Site	$4,695,000
C063. Construction—Attractions	$2,800,000
C064. Construction—Miscellaneous	$550,000
C071. Installed Equipment	$2,000,000
C081. Information Technology	$34,000
Total	**$11,361,150**
Profit (10%)	$1,087,100
Grand Total	**$12,448,250**

Appendix C: Risk Assessment

There is a separate Risk Management Plan document provided. A management overview of that document is shown in Table 3.5.

Table 3.5: Risk Evaluation Criteria

Probability Criteria	
Probability Rank	***Description***
High	Greater than 50% probability of occurring
Medium	Between 25 and 50% probability of occurring
Low	Less than 25% probability of occurring
Impact Criteria	
Schedule Impact	
Impact Rank	***Description***
High	Delay the opening of the park beyond July 1, 2005 (more than 6 weeks)*
Medium	Delay the opening of the park between June 15, 2005, and July 1, 2005 (3 to 6 weeks)
Low	Delay the opening of the park between May 24, 2005, and June 15, 2005 (less than 3 weeks)
Cost Impact	
Impact Rank	***Description***
High	Could add more than 20% to cost of the project (more than $2,500,000)
Medium	Could add 10 to 20% to the cost of the project (between $1,200,000 and $2,500,000)
Low	Could add less than 10% to the cost of the project (less than $1,200,000)

*Opening after the July 1, 2005, date is economically unsound, and delays beyond this date probably will result in deferring the opening of the Water Park to the following year.

Evaluation of Overall Project Risk

An evaluation of the overall project risk is given in Tables 3.6 and 3.7.

Table 3.6: Risk Table at Start of Project (Prior to Phase 1)

Probability *Impact*	*LOW*	*MEDIUM*	*HIGH*
HIGH	7. Construction company has major financial difficulties. 8. Major investor withdraws from project.		1. Failure to secure investors. 2. Political groups delay zoning approval.
MEDIUM		5. Unable to find suitable operations manager.	
LOW	4. Delivery of attractions delayed. 6. Design not approved. 9. Delays in obtaining construction permit. 10. Construction delays due to inclement weather.	11. Post-trial modifications delay opening.	

Table 3.7: Risk Table at Start of Phase 3

Probability *Impact*	*LOW*	*MEDIUM*	*HIGH*
HIGH	7. Construction company has major financial difficulties. 8. Major investor withdraws from project.		
MEDIUM		5. Unable to find suitable operations manager.	
LOW	4. Delivery of attractions delayed. 6. Design not approved. 9. Delays in obtaining construction permit. 10. Construction delays due to inclement weather.	11. Post-trial modifications delay opening.	

Risk Management Summary

A summary of the risk management is given in Table 3.8.

Table 3.8: Water Park Risk Management Summary
Revised October 8, 2002

Risk	*OPI*	*Probability*	*Impact*	*Status*
1. Funding not secured	TL (Finance)*	High	High	Open
2. Political groups delay approval	PM	High	High	Open
3. Unfavorable environmental assessment	PM	Medium	High	Open
4. Delay in delivery of attractions	PL (Design/Construction)	Low	Low	Open
5. Unable to hire suitable operations manager	PM	Medium	Medium	Open
6. Final design not acceptable	PL (Design/Construction)	Low	Low	Open
7. Financial difficulties with construction company	PL (Design/Construction)	Low	High	Open
8. Major investor withdraws	TL (Finance)	Low	High	Open
9. Delays associated with construction permits	PL (Design/Construction)	Low	Low	Open
10. Inclement weather delays construction	PL (Design/Construction)	Low	Low	Open
11. Post trial modifications exceed three-week window	PL (Design/Construction)	Medium	Low	Open

*TL: Team Leader; PM: Project Manager; PL: Project Leader.

Appendix D: Conceptual Design

Preliminary drawings of the Water Park, slides, parking area, roads, and main buildings are not included in this example.

4

Proposal

Request for Proposal: Discussion

A request for proposal (RFP) is a description of the requirements for either a project or a portion of a project that is to be contracted out. It is basically a call for proposals or bids to be made on the contract. The author of the RFP then will receive bids, do evaluations of all the proposals, and award the contract to the most deserving contractor.

Internal projects do not produce a RFP, but they may have a Requirements Document (RD) that details the client's description of a problem that requires a solution. The internal developers will develop the project as the solution to these requirements. The format of the RD can be similar to that of the RFP presented here, except that there will be no mention of competition.

Request for Proposal Outline

Cover Letter

Addressed to the prospective bidder or bidders, this letter outlines the most important information pertaining to the proposal: the return address, the deadline date, and how questions will be addressed.

Appendix A: Offer of Services

The logistics of the proposal is covered here, including the number of copies to be submitted and whether you wish to use the two-envelope system. The requirement for fairness in government contracts forbids the reader of the proposal from seeing the cost until the proposal is evaluated to meet the mandatory requirements. For this reason, the cost bid is placed in one envelope, and everything else in another.

Appendix B: Terms of Reference

This appendix is the meat and potatoes of the RFP. It consists of the following 12 sections.

Background Information

Identify the company (user) and the vendor(s) to whom the RFP is targeted. State the problems that need to be fixed, the history, examples of the problem situation, motivation to fix it, and so forth. This section is used to introduce the potential project developer to the buyer company or department if necessary, describing its culture, its environment, or the way it does business. It gives the potential project team a feel for the user and his or her problem. This section may include a summary of the basic deliverables required.

Purpose

This is the basic purpose of the solicitation and the end point.

Contractor's Role

Outline of the responsibilities of the contractor.

Deliverables Required

Provide a detailed list of required deliverables.

To Be Supplied by Buyer

Items that the buyer will furnish. This begins with the required financial support (progress payments), liaison, or other actions as required.

Other Details

As required, the location of the project or deliverables.

Delivery Date

Statement of the required date for submission of the proposal. Emphasize that this is firm or state that it is just a target date.

Replacement of Contractor Resources

Terms and conditions of what happens if the contractor cannot provide the resources promised in the proposal.

Suitability of Services

Terms and conditions of how the buyer intends to accept the deliverables.

Proposal Contents

A listing of the minimum proposal contents, which the buyer will use to evaluate the proposals. Also see Appendix C, "Evaluation Method and Criteria."

Basis of Payment

Whether the bid must be fixed price or on a cost and materials basis. Mention whether milestone payments are acceptable.

Assignment

Whether the buyer allows subcontracting of all or parts of the work.

Appendix C: Evaluation Method and Criteria

Add as an appendix the evaluation criteria that will be used to compare the proposals and select the winner. This includes the steps used in the selection process, the technical criteria considered in the scoring, and any caveats that may apply.

The proposal evaluation plan states how the buyer will impartially evaluate several proposals to choose the best solution for the project. The evaluation is based on certain criteria. These criteria are listed, and a weight is assigned to each (usually from most critical to least critical). Some criteria may be marked "mandatory." If these criteria are not met, the proposal is immediately disqualified. The various vendors that respond with proposals are scored on how well they meet the criteria.

Some contracting agencies, for example, the federal government, must score proposals without taking cost into account. These agencies may require that the cost be submitted in a separate envelope; this is called a "two-envelope" proposal. After the bids are scored, the cost envelopes for the compliant ones are opened and a cost per point is calculated. The team with the lowest cost per point and an adequately high total score should win the contract, but not necessarily.

Proposal Evaluation Plan

The proposal evaluation plan is published with the request for proposals to inform the prospective bidders about how they will be evaluated.

Proposal Evaluation Plan Outline

Selection Approach

In this section you list the steps that will be taken to evaluate the proposals. Here is where you would outline the two-envelope selection method, if applicable. We will see more detail on the proposal evaluation methodology later in this chapter.)

Selection Criteria

In this section you can detail the criteria that will be used to evaluate the proposals. The criteria can be general, such as experience required or management methodology required. Additionally, there may be technical criteria. This usually is done for products that have specific technical criteria; for example, the proposed computer hardware must have at least *x* bytes of RAM, *y* bytes of hard disk space, and so forth.

An example of an RFP follows. The format of this example is taken from a federal government solicitation. This format can be simplified for a smaller or less complex project environment.

Request for Proposal: Water Park Planning and Construction

Checklist of Documents

Invitation to Tender

Offer of Services: Appendix A

Terms of Reference: Appendix B

Evaluation Criteria: Appendix C

Invitation to Tender

File: WP-004
August 1, 2002
Subject: **Request for Proposal# WP-004**
Water Park Project Management

Dear Sir/Madam:

Carlington Aquatic Parks has a requirement to establish a contract to undertake the above referenced project in accordance with the Terms of Reference attached hereto as Appendix B.

If you are interested in undertaking this project, you are requested to submit a proposal in FOUR (4) copies, clearly indicating on the envelope or package "**BID/PROPOSAL WP-004**" together with the title of the work name and address of your firm, addressed to:

Carlington Aquatic Parks
123 Address Road
Ottawa, ON,
K2L 3J8

Proposals must be received at the above noted **address no later than September 1, 2002**. *Proposals received after September 1 will be rejected and returned to the sender unopened.*

Proposals submitted by **fax, e-mail,** or **Internet** will not be accepted.

No interpretation of the meaning or intent of the Request for Proposal (RFP) documents, nor correction of any apparent ambiguity, inconsistency, or error therein will be made to any tenderer orally. Such questions **must be in writing** and sent to Dan Milks, Carlington Aquatic Parks, fax (613) 123-4567, and must be received before **August 21, 2002**. All answers will be in the form of written addenda to the RFP and will be sent to all prospective tenderers.

Appendix A: Offer of Services

Proposals are to be submitted in four (4) copies together with two (2) copies of the completed Offer of Services (Appendix A) duly signed by the individual in the case of sole proprietorship, one or more partners in the case of partnership, or an authorized representative in the case of a corporation.

Proposals are to be submitted according to a two-envelope system:

Envelope 1: Technical Proposal (4 copies)

Your proposal is required to form the basis of a contractual agreement and should respond to all the requirements in the Terms of Reference and as detailed in Appendix C, the Evaluation Criteria. Your proposal is to be in sufficient detail to enable evaluation.

Note: **No cost information is to be included in Envelope 1.**

Envelope 2: Cost Proposal (2 copies)

Contractors shall complete and return two (2) copies of the Offer of Services Form in Envelope 2.

Note: Only cost information shall be provided in Envelope 2. All technical information supporting the proposal shall be in Envelope 1. Envelope 2 will be opened only after the technical evaluation is complete and the proposed resource has met the Mandatory Requirements and has achieved a minimum score of 70% of the total possible points on the Rated Requirements (Part A).

Both the Technical Proposal and the Cost Proposal envelopes are to be sealed and submitted together in a third envelope addressed to Carlington Aquatic Parks.

Proposals will be evaluated in accordance with the Evaluation Criteria attached hereto as Appendix C.

If additional information or clarification of any documentation is required, you are requested to contact the undersigned at (613) 555-1234 or by fax at (613) 123-4567.

The lowest cost, or any Proposal, will not necessarily be accepted.

Yours truly,

Dan Milks

President

Carlington Aquatic Parks

Kanata, Ontario

K2L 3J8

Appendix B: Terms of Reference

Carlington Aquatic Parks

Water Park Project Management, September 2002

Background Information

Carlington Aquatic Parks is a water park planning organization responsible for locating a project manager to oversee the construction of a water park in Nepean. This water park will have a capacity of 7000 persons per day. The proposed site is in the western portion of Ottawa–Carleton's greenbelt, within a 45-minute drive of over 1 million residents in eastern Ontario and western Quebec. The park will be serviced by local roads, public transportation, and bike trails.

1. The objective of this RFP is to find a project management firm to construct the Ottawa–Carleton Water Park by May 21, 2005.
2. The attractions will include: a children's wading pool, a standard swimming pool with diving boards, a wave pool, three water slides, and an artificial river ride.

Purpose

1. The purpose of this solicitation is to invite potential contractors, on a competitive basis, to submit a proposal to Carlington Aquatic Parks to establish a contract to provide the water park.
2. This contract is a one-time event, and upon successful testing of the finished water park, the contract will be deemed complete.

Contractor's Role

The contractor's role will be to provide the following:

1. A fully functioning water park
2. Project management plans and other documentation detailing construction of the water park
3. Financial plans
4. Risk Management plans and Quality Management plans
5. Contract Management documentation
6. Change Management documentation

Deliverables Required

The following items are to be provided by the contractor:

1. A fully functional, complete water park
2. Completed landscaping, parking lot, and paths
3. Upgraded traffic lights and roads
4. Full-time operations management team
5. Full-time and part-time human resources as park staff
6. Documentation relating to the operations of the water park
7. Support from the project team for several weeks after opening, until all testing and repairs have been completed

To Be Supplied by Carlington Aquatic Parks

Carlington Aquatic Parks will provide the following items:

1. Financial support (progress payments)
2. Liaison between project team and government organizations

Water Park Location

The Watt's Creek location and directions will be provided by Carlington Aquatic Parks on request.

Delivery Date

The water park must be completed and functional by May 21, 2005.

Replacement of Contractor Resources

The Contractor must provide the resources and facilities named in the contract, unless the Contractor is unable to do so for reasons beyond his or her control.

Should the Contractor at any time be unable to provide the facilities named in the contract, the Contractor shall be responsible for providing replacement resources and facilities at the same cost, of similar or better quality, and by the required date.

In advance of the date upon which the replacement resources and facilities will arrive, the Contractor shall notify in writing Carlington Aquatic Parks of the reason for the unavailability of the resources/facilities named in the contract.

The Contractor shall then provide to Carlington Aquatic Parks the specifications of the proposed replacement resources.

Under no circumstances shall the Contractor allow replacement resources and facilities that have not been authorized by Carlington Aquatic Parks.

Suitability of Services

All items and resources provided are subject to evaluation within a specified time from delivery, on the basis of quality and adherence to the Contractor's proposal. The items and resources provided by the Contractor must be of a level of quality acceptable to Carlington Aquatic Parks.

Should the items and resources from the Contractor be considered unsuitable and upon written notice from Carlington Aquatic Parks, the Contractor must provide suitable replacement items and resources in a timely manner.

Proposal Contents

Bidder's proposals will include the following when proposals are submitted:

1. An indication of the understanding of the requirement
2. A description of the methodology and approach to be followed
3. A summary outlining the bidder's expertise in performing similar service, including a list of references from former clients with contact names and phone numbers
4. A listing of proposed personnel, together with a statement of the appropriateness and quality of the proposed resource
5. Résumés clearly illustrating each individual's qualifications, knowledge, communication skills, and experience in relation to water park project management
6. A copy of the proposed water park project management plans

   Note: These plans will be returned to the unsuccessful bidders, and Carlington Aquatic Parks will not reproduce or make available the Bidder's plans to any other parties.
7. Costing elements as indicated in the Request For Proposal (RFP)
8. Any additional information the bidder considers necessary or wishes to include in the proposal

Basis of Payment

Bidders shall submit their tender price as an all-inclusive fixed price. Payments based on milestones may be negotiated at a later date.

Assignment

The Contractor shall not, without the written consent of Carlington Aquatic Parks, make any assignment of this agreement or any subcontract for the performance of the service hereby contracted for.

Table 4.1: Technical Selection Criteria

Proposal Item	*Weight*	*Score (1–10)*	*Total (Weight × Score)*
1. Previous relevant experience (management of planning and construction of water parks Can$8 million or more in value)	Mandatory		
2. Previous projects completed on time and within budget	15		
3. References from satisfied clients	10		
4. Financial stability	10		
5. Project team composition	8		
6. Deliverables	6		
7. Reliability and availability	6		
8. Use of technology	5		
9. Location of offices	5		
10. Change management	5		
11. Risk management	5		
12. Quality management	5		
13. Communications (interface to client)	5		
14. Contract conditions (fixed versus per diem payments, etc.)	5		
15. Proposal format and quality	5		
16. Warranty	5		
Grand Total	**100**		
Cost			
Cost/Point			

Appendix C: Evaluation Criteria

Request for Proposal: Water Park Project Management

Selection Approach

The selection process will follow the following four steps:

1. Determine the compliance of each bid with the requirements stated in the Terms of Reference.
2. Evaluate each proposal against the technical selection criteria, as shown in Table 4.1. Based on the evaluation, each proposal will be assigned a score. Bidders must achieve a minimum score of 70% of the points available for each element, and the overall score must achieve 75% of the total technical points to qualify for the next step.
3. Open the cost envelopes and compute the total cost/point of the qualified bids (the final evaluation phase).
4. Select the winning bid after consideration of the results of the entire selection process.

Note: Carlington Aquatic Parks is not obliged to award a contract to any of the submitted bids.

It is essential that the elements contained in your bid be stated in a clear and concise manner. Failure to provide complete information will be to your disadvantage.

Proposal: Discussion

The proposal is the project team's answer to the Request for Proposal. It is a more precise definition of the project. It presents the initial cost estimate, although this may be up to 50% over or under the actual cost of the project. Its main purpose is to gain approval of funds and commitment of stakeholders.

The proposal is similar to a project charter; however, the charter is an internal document, whereas the proposal is external. Just as the proposal acts as a contract between the buyer and the (external) vendor, the charter is the commitment of the (internal) developers.

The proposal should be as short as possible. It should be written by the Project Manager (PM), with technical advice provided by the Project Leader and technical experts. At times the marketing organization writes the proposal, but the PM should provide input.

Proposal Outline

The contents of a proposal generally include the following 15 items.

Cover Letter

This is a letter written to the decision maker that is signed by the Project Manager; if an account representative is the primary contact with the client, he or she may sign as well. The body begins with introductory text such as "Thank you for giving KLSJ Consulting the opportunity to propose the development of a water park."

The next paragraph should give a simple description of the major deliverables, such as "capacity of 7000 guests per day . . . unique attractions including an outdoor wave pool, artificial river, and water slides."

In the subsequent paragraphs, explain if this is a proposal for everything asked for in the RFP or a subset only. State price and delivery date if it is a "fixed-price" contract and cost per hour and estimated hours if it is a "cost-plus" contract. A cost-plus contract specifies you will work *x* hours and be paid by the hour, plus materials used.

Close the sale: End the letter with statements that force a quick decision in your favor. You do not wish to wait six months to get an answer. Closing remarks could be expiration dates for the price quoted (30 days is customary) or statements such as "if we are given a go-ahead in 14 days, we can start January 1; otherwise we must do another project, and we can only start yours on . . . "

Title Page

This page is titled "Proposal" and contains the title of the project, author, date, revision number, company logo, and so forth.

Table of Contents

Since the client may not be familiar with your proposal format, give a brief explanation of the purpose of each section. Here is an example of the format:

Section 1: Scope

Describes the business problems addressed by the KLSJ solution, the size, extents, and limits of the proposed park.

Try to follow the outline of the RFP: The closer in structure it is to the scoring key, the easier it will be for the client to evaluate it.

The remainder of the proposal contains the sections that follow:

Scope

See the paragraph above. This section also could be called "objectives," because it lists the major deliverables. Much of this section comes directly from the requirements document. Basically, you want to show that your firm can meet the requirements of the client.

Advantages

Sell the Project Team here. Prove how your well-planned, well-controlled methodology will work. Address any requests in the RFP that your team is particularly adept at providing. You may even consider a few remarks about your competition's (inferior) solutions. Be careful not to make explicitly disparaging remarks about the competition.

Financial

Note that this will be kept in a separate section under the two-envelope system.

State the total price and the delivery date. If it is a fixed-price contract, quote only the final price. Although the cost estimates may come from a much more detailed calculation, do not divulge those details because you do not wish to let them get into the hands of the competition. If you are proposing a cost-plus contract, state a nonbinding estimate of the hours or days required and the cost per hour or day.

State the expected delivery date (assuming you are given a prompt go-ahead).

Draw a payback graph that shows total cost, the cost savings or benefits of the project once it is built, and the number of months or years it will take for the project to pay for itself.

List any nonmonetary benefits: job satisfaction, goodwill, customer happiness, management happiness, and so on.

Quality

In a small project, quality management is addressed as part of the total plan. In a larger project, a separate quality management plan is written. Here you should explain what systems you will be using for quality planning, quality assurance, and quality control. Most important, impress on the client that both the resulting products and the project management processes will be under the strictest quality control.

Risk

In a small project, risk management is addressed as a part of the total plan. In a larger project, a separate risk management plan is written. Your risk management strategy should be explained here. Show that you did an initial risk assessment of the project. Explain what the major risks to the project are and how they can affect project costs. Most important, prove to the client that you will manage project risks formally.

Communication

The communications plan you choose to use should be explained here. You should state who the major stakeholders in the project are and the steps that will be used to communicate with them. Explain *who* must receive *what* information, *why* and *when*, and *how* it will be communicated. A format such as the one shown in Table 4.2 can be used.

Project Plan

Describe the steps you have planned to develop the project.

Describe the project organization. Describe the milestones, especially the ones the client will have an opportunity to review. Show how the client will be informed about project progress.

List the responsibilities the client will have. This may be people or materials that he or she must provide or a task he or she must perform. The client must appoint a project representative who can answer your questions. He or she also must provide approvals of the signed-off documents.

Table 4.2: Communications Plan for Major Stakeholders

Who	*What*	*When*	*Where/How*	*Why*
PM and Client	Costs	Phase 1	Meeting	Approval
All lines of communication are listed				

The Proposal will be signed by both sides. Make the client's responsibilities as clear as possible.

Describe the activities you will do in each phase. Show that you know what you are doing (remember that this is a sales document).

Deliverables

List what the client will receive: equipment, facilities, human resources, and so forth.

Warranties. State how long after delivery you intend to fix any problems and how you will provide support.

Documents. List the manuals that will be provided (operating, management, maintenance) with a brief description of the purpose and the intended reader.

Training. List the training courses that will be provided (operator, manager, maintenance person) with a brief description of the purpose and the audience.

Delivery. State when you will deliver and how that will be done.

Acceptance

You must implement an acceptance method that proves unequivocally that the project team has met its commitment. Usually this is a formal demonstration of every function that the system will accomplish. Describe the method of demonstration.

Alternatives

Sometimes you will find that the RFP has been written with a certain vendor in mind. This is fine if you are that vendor, but what if you are not? You then must detail the other vendor's solution as an *alternative* solution and prove why your solution is better. Emphasize your team's advantages as detailed in "Advantages."

Terms, Conditions, and Assumptions

List here all the conditions you want to work under. They are listed here to protect you. Be especially careful with cost-plus projects: Promise to deliver only the hours, not products.

List all assumptions. There will always be questions that the client was unable to answer precisely, and so the answers were assumed. If these assumptions affect the cost of the project, you must protect yourself.

Terminology, Glossary, and Acronyms

Even though the Proposal must be written using the client's language as much as possible, some technical terms may have crept in. If you feel these terms are unfamiliar to the client, define them here.

Proposal: Example

KLSJ Consulting

14 Palsen St.,
Ottawa, ON,
K2G 2V8

June 1, 2002

Attn: Mr. Dan Milks
President
Carlington Aquatic Parks
123 Address Road
Ottawa, ON,
K2L 3J8

Dear Mr. Milks:

Thank you for giving KLSJ Consulting the opportunity to propose to you the planning and construction of the Ottawa-Carleton Water Park. Please find enclosed our proposal. As requested, a two-envelope system has been utilized, with technical information in Envelope 1 and financial and cost information in Envelope 2.

KLSJ Consulting is proposing the design, development, and construction of the park. Prior to the official opening we will hand over operations to the operations management team, which will manage the park. The Water Park is to be situated on land owned by the National Capital Commission in the "greenbelt" area of western Ottawa. The project will include the planning and construction of various attractions, including slides, an action river, a wave pool, swimming pools, and some dry attractions such as a tennis court and a rock-climbing wall.

The Water Park we propose will have a capacity of 7000 guests per day. It will include attractions such as an outdoor wave pool, an artificial river, and water slides. The park will be serviced by local roads, public transportation, and bike trails. The parking area will be designed to accommodate 1000 cars.

This is a proposal for the design and building of the project, with an expected delivery date of

May 2005. Project duration is estimated to be from September 2002 to May 2005, a period of 32 months. The project risk is expected to be high for Phases 1 (Concept) and 2 (Planning), but once the issues of land approval and capital investment have been addressed, the project risk will become low for Phase 3 (Execution) and beyond.

If KLSJ is given a go-ahead by September 1, 2002, we would be able to start your project at once and complete it by May 2005. If project approval is not received by September 1, we will not be able to begin the Ottawa-Carleton Water Park project until early in 2003, which could compromise the May 2005 target completion date.

We are looking forward to receiving a response to this proposal in a short while and working with Carlington Aquatic Parks.

Sincerely,

Karen Dhanraj
Project Manager
KLSJ Consulting

KLSJ Consulting

June 1, 2002

Project Proposal

Ottawa–Carleton

Water Park

14 Palsen St., Ottawa, ON, Canada, K2G 2V8

Envelope 1: Technical Information

Contents

Section 1: Scope

Describes the business problems addressed by the KLSJ solution: the size, extents, and limits of the proposed park.

Section 2: Advantages of KLSJ

Explains why KLSJ is the right choice for this project.

Section 3: Quality Management

Explains systems to be used for quality planning, quality assurance, and quality control.

Section 4: Risk Management

Explains risk management strategy and initial risk assessment.

Section 5: Communications Plan

Explains how KLSJ will interface with CAP and other stakeholders.

Section 6: Project Plan

Describes the Project Team. Explains client involvement. Describes how the project will be run, including the major phases, activities, and milestones.

Section 7: Deliverables

Explains what the client will receive.

Section 8: Acceptance

Explains the criteria for acceptance of the project by the client.

Section 9: Alternatives

Explains alternatives to the proposed solution.

Section 10: Terms, Conditions, and Assumptions

Explains the conditions on which the project team will work; explains assumptions made by the Project Team.

Section 1: Scope

The overall objective of this proposal is to describe how KLSJ Consulting will complete the design and construction of the Ottawa–Carleton Water Park facility on behalf of Carlington Aquatic Parks. The Water Park is to be completed and ready for a scheduled opening of May 21, 2005.

The responsibility of KLSJ Consulting is to facilitate the creation of a top-quality recreational facility (see Appendix A, Preliminary Drawings) with the following criteria:

- Capacity of 7000 guests per day

---

*[Author's note: The final document should include illustrations in this appendix. For the sake of expediency, illustrations do not appear here.]

- Unique attractions, including an outdoor wave pool, an artificial river, and water slides (see Appendix A, Preliminary Drawings)
- A parking area adjacent to the Water Park capable of accommodating 1000 cars (see Appendix A).

We note that the proposed site is in the western portion of Ottawa's greenbelt, within a 45-minute drive of over 1 million residents, and that it is currently served by local roads, bus transportation links, and bicycle trails. The proposal assumes that roads, utilities, and services up to the site perimeter will be the responsibility of local governments and are beyond the scope of this project.

Given the fixed target date for the opening of the park and the low probability that extra funding could be located on short notice, any project related problem is likely to result in a decrease in the number of attractions or their quality. KLSJ Consulting can provide and execute an effective project management system that will ensure that the intended cost, time, and quality are protected.

Section 2: Advantages of KLSJ

The major advantage KLSJ has over our competition is the proposed project team. The team consists of five highly qualified core members (see below). The five team members have over 50 years of combined experience on related projects. The team has worked together on five similar projects in the past. The team and their past projects include the following people:

Karen Dhanraj (Project Manager)

- Toronto's Skydome, Paramount's Wonderland Water Attractions, Jetform Ball Park, and Capital Racing Go-Karts.

Laverne Fleck (Team Leader, Legal)

- Paramount Wonderland Water Attractions, Toronto's Skydome, Seattle Water Park

Steve Jackson (Team Leader, Finance)

- West Edmonton Mall, Seattle Water Park, Calgary Stampede Grounds

Scott Kennedy (Team Leader, Design and Construction)

- Paramount Wonderland Water Attractions, West Edmonton Mall, Famous Players Coliseum movie theaters in Washington, DC

Jim Harris (Risk Manager)

- Six Flags Amusement Park in California, West Edmonton Mall, Seattle Water Park

Our combined years of experience and our comfortable working relationship will guarantee the success of this project. We are particularly skilled at quality management, risk management, and construction management.

In all our previous projects, the level of customer satisfaction was very high upon project delivery (see References, Appendix B).

Section 3: Quality Management

Quality Planning

The Quality Plan is a key document for highlighting and clarifying quality issues and expectations by the client. Quality planning will involve input from both KLSJ and the client, after which the plan will be approved by the client. The objective will be to create a sound plan that will enable the client and the Project Team to set quality standards collectively for both the product and the management process. The quality standards will have to be very high for the Water Park project, since human safety definitely will be an issue. KLSJ has always provided exceptional quality planning for all its projects.

The relationship among time, cost, and quality (the three major constraints on any project) will be taken into consideration. The need for rigid quality standards is essential for this project, but that could mean an increase in costs or a time delay in order to complete the Water Park with the appropriate quality standards.

Quality Assurance

Quality assurance is intended to provide the client with confidence that if the plan is implemented properly, the project will satisfy the specified quality standards and also meet the client's stated needs. It ensures that the Quality Plan will be carried out. The Quality Plan will assign responsibility for each major aspect of quality to a specific person or position within the project team. These individuals will be part of the quality assurance system.

Quality Control

Quality control involves monitoring the actual processes, as opposed to quality assurance, which involves monitoring implementation of the plan. It is the activity (monitoring, inspecting, testing, etc.) necessary to ensure that products and services meet the required quality level. To maintain quality control, it may be necessary for KLSJ to purchase special testing and inspection equipment for the Water Park.

Section 4: Risk Management

The risk management strategy used by KLSJ involves several steps:

1. Complete a thorough list of the possible risks to the project.
2. Quantify the probability of these risks and the impact they would have on the schedule, budget, or quality of the project. (The project team will do this.) This will result in a risk table

Table 4.3: Risk Table

Probability ***Impact***	***LOW***	***MEDIUM***	***HIGH***
HIGH	7. Construction company has major financial difficulties. 8. Major investor withdraws from project.	3. Environmental assessment requires major mitigation.	1. Failure to secure investors. 2. Political groups delay zoning approval.
MEDIUM		5. Unable to find suitable operations manager.	
LOW	4. Delivery of attractions is delayed. 6. Design is not approved. 9. Delays in obtaining construction permit. 10. Construction delays due to inclement weather.	11. Post-trial modifications delay opening.	

such as the one shown in Table 4.3. This table will tell us (and the client) which risks have the highest priority.

3. Formulate a mitigation strategy for the highest priority risks. The results of this thorough analysis will be compiled in a risk log.

4. Monitor and control the risks as the project progresses. If the anticipated risks materialize, a decision can be made whether to implement the planned mitigation strategy. If unforeseen problems are encountered, they will be analyzed and a response will be developed as necessary. The status of all the risks, anticipated or not, is updated in the risk log.

Following is the initial risk assessment for the Ottawa–Carleton Water Park:

- Total project risk classification: high
- Uncertainty with investors and funding as well as potential difficulty in obtaining approval for the land and zoning make this a high-risk project.

- Once the initial approvals and funding have been obtained, the project risk for the construction of the water park will be reduced to low.

Section 5: Communications Plan

The major stakeholders in the Water Park project must be communicated with on a regular basis. Table 4.4 [see pages 82 and 83] identifies *Who* we will communicate with; *What* information will be given; *When*, *Where*, and what method of communication will be used; and the purpose of the communication (*Why*).

Section 6: Project Plan

KLSJ has a five-phase planning approach based on the project management standards documented by the Project Management Institute's *A Guide to the Project Management Body of Knowledge (PMBOK Guide)*. The five phases of the Project Plan are as follows:

- *Phase 1: Initiation.* This phase involves developing and assessing the feasibility of the overall project concept. Key activities include the owners' partnership agreement and the initial feasibility study.
- *Phase 2: Planning.* This phase involves the development of plans and strategies required in managing the project. Key activities include creating the business plan and project plan, completing preconstruction (engineering) studies, and locating investors.
- *Phase 3: Execution.* This phase involves the actual construction of the facility and also includes an increased focus on marketing.
- *Phase 4: Handover to Operations.* This phase involves the process of finalizing the facility, preparing it to open and turning over operational control to the park manager.
- *Phase 5: Closing.* This involves cleanup, final payments, the post-project report, and the final report to the client.

Table 4.5 [see pages 84 and 85] shows the key milestones on the critical path within each of the phases.

In addition, there are other important milestones that are not part of the critical path, yet are essential for schedule progress. These are shown in Table 4.6 [see page 86].

Changes to the Plan

KLSJ will be responsible for tight schedule control. There is a specific deadline for each phase and the total project. Changes can delay these deadlines, add to project cost, and affect project quality. Hence, all changes must be handled by using a formal change management process. Any stakeholder, client, or contractor will be required to fill in a Change Request Form (to be provided). KLSJ will analyze the impact of all changes. If the requested change does not disrupt major deadlines or add significantly to the cost of the project, it may be implemented. If, however, the requested change will delay the project or add to the cost, KLSJ will document this on the Change Request Form. The client must sign off on these changes, thereby accepting possible project delay and/or cost increase.

Progress Reporting

To provide project control, each phase concludes with an end-of-phase review and approval milestone. In addition to these built-in schedule controls, KLSJ also will use the following:

- Performance Reports at the end of each phase to provide an overview of the progress of the project during that phase, including schedule, cost, and quality of the product. Significant change decisions made during the phase also will be included in this report.
- Quarterly Status Reports will be provided to all key stakeholders, in addition to a Biweekly Status Report provided to the client. The format of these two status reports will be similar, and is as outlined below:

  From: Project Manager

  To: Key stakeholders/Client, Project File

  Reporting period:

  Date:
  - Background
  - Activities and accomplishments this period
  - Problems encountered
  - Problems solved
  - Problems still outstanding
  - Schedule progress versus plan and trends
  - Expenses versus budget
  - Plan for next period

In addition, the Status Reports will contain a tracking Gantt Chart produced using Microsoft Project software, showing schedule progress compared to the baseline plan.

Section 7: Deliverables

KLSJ Consulting will deliver the following:

- A fully functional, complete water park
- Completed landscaping, parking lot, and paths

Table 4.4: Stakeholder Communication Chart

Who	*What*	*When*	*Where*	*Why*
Project Team (PT) with Client (Carlington Aquatic Parks)	Report progress	Biweekly report (written) Weekly updates: telephone, e-mail, informal meeting		To keep the client apprised of changes and delays and to warn client if there are significant changes
	Significant problems encountered	As they arise: informal or formal meeting as appropriate	PM office or client office as convenient	To discuss options, get decisions, and solicit assistance
PT with Investors	Report progress	Continuously	Internet site	To apprise of progress (avoid rumors arising from inaccurate media reports); general information
		Quarterly/end of phase progress reports (written documents)		
PT with general public	Press releases News articles	Major milestones/events	Newspapers, radio, local magazines, etc.	To keep all informed, garner interest, marketing, gain support when hindered by political issues
	Progress/general information	Continuously	Internet site	

PT with Community groups	General information, project intentions (open forum meetings, general (presentations)	Prior to seeking government/regulatory approvals	At public meeting place	To alleviate fears about effects on the neighborhood, traffic, environment, etc.
		If there is a potential conflict		To keep all informed of process
PM with Government agencies (MCVA, NCC, City of Nepean)	Grease the wheels (informal meetings, consultations)	As often as practicable prior to seeking approvals formally.	Officials' offices, business luncheons (general schmoozing)	To ease approval process, identify potential problem areas, garner support
	Seek approvals (formal submissions, presentations)	After necessary studies and consultation completed	Agency offices, meeting rooms	To obtain formal regulatory approvals
PM with Contractors (various study consultants, marketing firm)	Inform of upcoming work/project	Prior to release of RFP	Phone calls, e-mail to offices	To solicit interest in submitting a bid
	RFPs Walk-throughs	As per work schedule Prior to bid submissions	Formal tender Site location	To gather bids To describe work, clarify issues
	Biweekly progress meetings	While performing work	PM office	To update progress, voice concerns
Project Team with each other	Report progress Coordination Problems/issues	Constantly (e-mail, telephone calls, informal meetings, memos) Biweekly meetings	PM office	To keep everyone informed of progress and problems encountered, coordinate work

Table 4.5: Key Milestones on Critical Path

Milestone	*Date*	*Key Stakeholder*	*Key Trigger Event(s)*
Phase 1: Concept			
Approval of project charter/partnership	November 2002	Project Manager	Creation of core project team
Selection of preliminary Water Park site	January 2003	Project Manager	Assessment of site options
Approval of Phase 1 and end of phase review	March 2003	Project Manager	Presentation to NCC authorities
Phase 2: Planning			
Approval of project plan	April 2003	Project Manager	Approval of Phase 1 (Concept)
Zoning approval from city	February 2004	Project Manager	Conduct environmental assessment, site approval
Approval of Phase 2 and end of phase review	February 2004	Project Manager	Zoning approval, confirmation of investors

Phase 3: Execution			
Approval of construction	November 2004	Project Leader	Completion of construction, installation of attractions, landscaping
Approval of Phase 3 and end of phase review	November 2004	Project Manager	Approval of construction
Phase 4: Handover to Operations			
Trial opening date	April 2005	Project Manager	Final site cleanup and set-to-work preparations
Handover to the operating team	May 2005	Project Manager	Posttrial modifications
Grand opening ceremony	May 2005	Operations Team	Handover to operating team
Phase 5: Closing			
Completed postproject review	June 2005	Project Manager	Handover to operating team
Project completion	June 2005	Project Manager	Project presentation to client, financial reconciliation

Table 4.6: Important Milestones Not on the Critical Path

Milestone	*Date*	*Key Stakeholder*	*Key Trigger Event(s)*
Phase 1: Concept			
Signing the partnership agreements	November 2002	Team Leader (Legal)	Create core project team
Approval of the architectural design	June 2003	Project Leader	Approve Phase 1 (Concept)
Phase 2: Planning			
Investors' commitment	December 2003	Team Leader (Finance)	Approve detailed business plan
Site approval from the NCC	December 2003	Project Manager	Presentation to NCC

- Upgraded traffic lights and roads
- Full-time operations management team
- Full-time and part-time human resources (park staff)
- Documentation relating to the operations of the water park
- Support from the project team for several weeks after opening, until all testing and repairs have been completed

Specific documentation to be delivered includes the following:

- Feasibility Study
- Business Plan
- Project Plan and Project Budget
- Risk Management Plan
- Environmental Assessment
- Zoning Approval
- Architectural and Engineering Design
- Project Presentations (for political and community groups)
- Investor Commitments
- Acceptance Test Plan
- Marketing Plan
- Operations Staff Guidance (financial, marketing, operations)
- Postproject Final Report

Section 8: Acceptance

Acceptance will involve thorough testing and demonstration of the water park operations to the client before the actual delivery of the system. This demonstration will be a "scripted" item-by-item inspection and testing of every park element. This Acceptance Test Plan will be agreed to and signed by the client before the start of acceptance. If the client agrees that all the functions perform in accordance with specifications, the client will pay KLSJ Consulting all funds still owing. If they do not meet specifications, a 10% holdback fee will be acceptable until KLSJ corrects the nonconformance.

Section 9: Alternatives

There may be other bidders for management of the project. Beware of some local companies with limited experience. These companies may only have experience gained in the installation of public swimming pools in the local area. This project is of significantly greater complexity, requiring a project team with a much broader range of knowledge and background to ensure success. There are a very limited number of companies with this breadth of experience; KLSJ is one.

Section 10: Terms, Conditions, and Assumptions

The project plan is founded on the following *key assumptions*:

- Economic and political environmental factors will continue to support the project concept.
- Community support for construction and ongoing patronage will continue.
- Sufficient debt and equity investors can be located no later than mid-February 2004.
- The project start-up in September 2002 and opening in May 2005 are fixed milestones.
- A suitable location can be identified and acquired in a timely fashion.
- The environmental assessment will be favorable or mitigation can be made with reasonable effort.
- The investors will be satisfied with the engineering design.
- Contractors of sufficient size and expertise are available.
- Delivery of all equipment is within the time allotted.
- All permits and permissions are secured within the time allotted.
- The weather during Phase 3 and Phase 4 remains within normal seasonal parameters.

- A suitable operations manager at a reasonable salary can be found.
- The trial run reveals no major problems.

KLSJ believes that these assumptions are reasonable. If any of these assumptions prove wrong, there undoubtedly will be an impact on project cost, time, and/or quality.

Appendix A: Drawings

Any preliminary drawings and designs would appear here.

Appendix B: References

All references of personnel and resumes would appear here.

Envelope 2: Financial Information

The initial estimated cost of the project is roughly Can$11,784,500. These costs will need to be refined during detailed project planning. The planned completion date, given a go-ahead by September 1, 2002, is May 21, 2005.

The cost estimates can be broken down into the areas shown in Table 4.7.

To provide more visibility into our costing methodology, project cost estimates have been broken out by project category, as shown in Table 4.8.

Note that these cost estimates are Class B with a range of +25% to −10%.

Table 4.7: Cost Estimate by Project Phase

	Fixed Costs	*Resource Costs*	*Total Costs*
Totals by Project Phas*			
Phase 1. Concept	$25,000	$100,000	$125,000
Phase 2. Planning	$200,000	$175,000	$375,000
Phase 3. Execution	$10,000,000	$125,000	$10,125,000
Phase 4. Handover to Operations	$70,000	$30,000	$100,000
Phase 5. Closing	$0	$30,000	$30,000
Total	**$10,295,000**	**$460,000**	**$10,755,000**
Profit (10%)	$1,029,500		$1,029,500
Grand Total	**$11,324,500**	**$460,000**	**$11,784,500**

Table 4.8: Cost Estimates by Project Category

Totals by Project Category	*Fixed Costs*	*Resource Costs*	*Total Costs*
Project Management	$400,000	$200,000	$600,000
Contract Management	$0	$30,000	$30,000
Financing	$20,000	$80,000	$100,000
Political/Legal	$0	$25,000	$25,000
Construction	$9,600,000	$100,000	$9,700,000
Marketing	$275,000	$25,000	$300,000
Total	**$10,295,000**	**$460,000**	**$10,755,000**
Profit (10%)	$1,029,500		$1,029,500
Grand Total	**$11,324,500**	**$460,000**	**$11,784,500**

Proposal Evaluation: Discussion

We have discussed the Request for a Proposal, the Proposal, and the Proposal Evaluation Plan (part of RFP).

The selection of the winning proposal normally follows these steps:

1. Ensure that the bid has met all the predefined requirements and is compliant. You must ensure that the bid proposes to deliver the requirements stated in the Request for Proposal. Check that the basic format of the proposal is acceptable (the closer in structure it is to the RFP, the easier it is to evaluate), the terms and conditions are accepted, and items marked "mandatory" in the evaluation criteria are present. Disqualify any noncompliant bids.
2. Evaluate the proposal against the technical selection criteria published in the RFP. Based on the evaluation, assign a score to each evaluation item. Discard any bids that do not achieve the promulgated minimum score or will not meet the performance requirements. Inform the discarded bidders of the reason for disqualification.
3. Implement the final evaluation phase. This generally takes into account the results from the technical evaluation along with the overall cost of the proposal. The price is given a rating and included in the overall evaluation to compute the total cost per point of the proposal. If you are using a two-envelope evaluation, this is when the cost envelopes are opened.
4. Evaluate the bids not only on the number of points but also on the total score. Remember, the higher the score, the better the bid. Inform the losing bidders why they were chosen. Where possible, incorporate your own judgment, experience, and knowledge of the bidders when making the final decision.
5. Contact the winning bidder. If there are problems with the winning bidder, you still have the option of approaching the second-place firm.
6. Inform the losers that they were not selected. Explain why they were disqualified or why their bids were not chosen. Return their material if requested.

Note: You are not obliged to award a contract to any of the submitted bids. None of them may meet minimum requirements or budget constraints.

Proposal Evaluation Outline

Tables

For each compliant vendor, a table filled in with the required functions, weights, and scores is provided. The weighted scores are totaled, and cost per point is calculated.

Choice of Vendor

The justification for the choice of vendor is stated.

Proposal Evaluation: Example

Table 4.9: Technical Selection Table

Proposal item	*Weight (%)*	*Score (1 to 10)*	*Total (Weight × Score)*	*Comments*
1. Previous relevant experience (management of planning and construction of water parks Can$8 million or more in value)	Mandatory	✓	✓	Vendor A has most
2. Previous projects completed on time and within budget	15	9	135	
3. References from satisfied clients	10	8	80	Two out of three satisfied
4. Financial stability	10	8	80	10 years of experience
5. Project team composition	8	10	80	
6. Deliverables	6	10	60	
7. Reliability and availability	6	9	54	
8. Use of technology	5	8	40	
9. Location of offices	5	7	35	
10. Change management	5	8	40	
11. Risk management	5	8	40	
12. Quality management	5	8	40	
13. Communications (interface with client)	5	7	35	
14. Contract conditions (fixed versus per diem, payments, etc.)	5	8	40	
15. Proposal format and quality	5	10	50	
16. Warranty	5	8	40	
Grand Total	**100**		**849**	
Cost			$250,000	
Cost/Point			**$1,143**	

Table 4.10: Comparison of Vendors

Vendor	*Points*	*Cost/Point*
A	849	1,143
B	721	1,023
C	703	1,342

An example of a completed technical selection table for vendor A is shown in Table 4.9.

A similar table is filled in for all compliant vendors; in our case, it would be vendors B and C (see Table 4.10).

Choice of Vendor

Although vendor B has the lowest cost per point, vendor A may be chosen since it has the most experience, the highest point score and is still within the budget constraints of the buyer.

Planning Phase Documents

5

Project Plan

Project Plan: Discussion

A project plan is intended mainly for the eyes of the developers. It contains a detailed step-by-step explanation of how the project will be built. In fact, it may be unwise to show it to an external client, especially in a competitive environment. We have seen cases where the client gave the plan to the favorite contractor or even did the project himself or herself. In this case, you should develop a project proposal (see Chapter 4) as a subset of the plan. If the contract already has been signed or it is an internal project, the Project Plan may be given to the client and used for final project approval.

Project Plan Outline

Executive Summary

This is an optional section that summarizes the major points of the document. Give the readers a chance to decide whether they need to read the remainder.

Background

Provide the readers a little history; they may not know anything about the project. State here how the project started, why, the original high-level objectives, a ballpark cost, and the schedule. Some or all of this information may be contained in the Project Concept (Chapter 1) or Project Charter (Chapter 3).

List all the external stakeholders and groups involved in the project. These are groups that provide information or resources, equipment, or advice: anyone the project is dependent on. Detail their relationship to the team, their responsibility (what they are to provide and when), their function, and their interfaces with your team. Be especially careful with customer-furnished equipment. Most important, state the impact if any of these groups does not deliver on time. All these stakeholders will read this plan, and you must hope they will be committed to their tasks.

Project Objectives

Objectives should be clear, tangible, specific, simple, realistic, and attainable.

In this section, detail the objectives stated broadly in the background section. List the major deliverables to be produced, both product and process. High-level functionality and performance may be mentioned here, as well as major constraints to be overcome. The ballpark cost/time and the delivery date are given here.

Make sure any objectives can be proved as having been attained at acceptance time. Also ensure that the objectives are realistic. This is especially important if some of the stakeholders are trying to impose an unrealistic deadline or cost. This is the time to bring expectations down to earth.

The Project Team

General

List the names of the individuals who are known to you. If the actual names are not known yet, state the knowledge level required for each position. In large projects, common position descriptions may suffice. It is important to describe skills requirements because there can be a large productivity difference between suitable and unsuitable project staff.

Organization

Diagram the organization of the Project Team, as shown in Figure 5.1.

Responsibilities

Detail the responsibilities of each member, as in this example:

- *Team members*. Responsible for the development work, reporting progress.

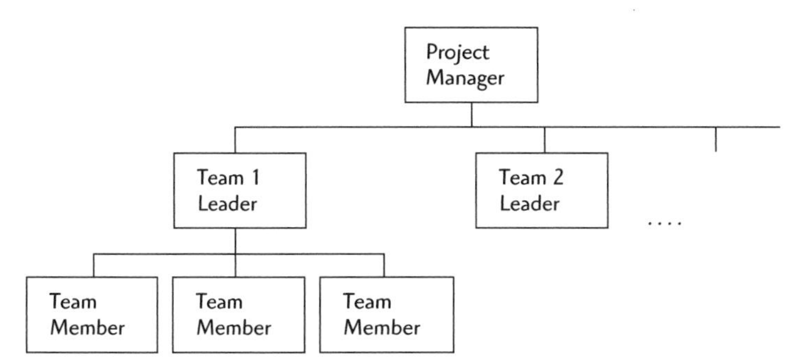

Figure 5.1 Project Team organization.

- *Team leaders.* Supervise team members on technical details only. Responsible for managing, but not necessarily doing, most complex activities. Help any team member with technical problems. Their major focus is the technical quality of the product.
- *Project manager.* Manages the team (leader, motivator, etc.). Responsible for all outside communications (reporting, meetings, client and upper-level management interface). Responsible for planning and controlling the project. Manages the human resources, including performance appraisals. The major focus is the successful completion of the project (on time, on budget, and with a product satisfactory to the client).

Authority Relationships

Explain the relationships in the organization chart: who reports to whom. Also note solid-line (direct supervision) versus dotted-line authority (advisory only).

Project Management Strategy and Methodology

Here you can detail the methods, approaches, shortcuts, and strategies you will use: Will all or part of the work be contracted? What alternatives were considered, and why was a specific one chosen? Are there major implementation shortcuts? Detail how the project will be managed; for example, there could be a Project Manager for both the buyer and the contractor who will define a division of responsibilities. Detail the project management methodology (*PMBOK* or other references) or other management standards used (e.g., ISO 9004-6). Inform the reader about the project management software tools that will be used (such as Microsoft Project).

Project Scope

Include the Work Breakdown Structure (WBS) at a high level only. Be careful here: If the competition gets your detailed plan, you will lose a major advantage. Explain the logic behind the breakdown: phased by function or time, process elements including inputs, analysis, implementation, tracking, and configuration management.

Project Schedule

As before, keep it at a high level. Explain the major dependencies, especially the external ones over which you have little control. How was the schedule it developed? Forward, or backward from a deadline?

Detail how you will control the schedule. Perhaps state that you will accept changes but use standard methods for change management.

List the major milestones in the schedule, their meaning, and implied percentage completion. State if there is any involvement of stakeholders, for example, approval of a deliverable or a milestone meeting.

Project Cost

Again, as above, at a high level only. State the accuracy or range of the estimate. Explain how it was calculated. Detail how you will control the costs.

Resource Estimates and Leveling

List the people you are expecting to work on the project as well as other resources required, such as equipment. Show where the schedule was adjusted and the workload was leveled—where some items had to be rescheduled owing to resource constraints. State whether the project could be sped up if more resources become available.

Quality Control and Performance Measures

Here you can list the standards you will use for the management of quality (e.g., ISO 9001–2000). State the commitment of the total project team to quality. List the actions you will take to ensure quality not only of the deliverables but of the management of the project. Show how you will assure that this quality plan is met. Most important, state how you will control the quality of the project: the quality of the processes using scope, cost, and time progress control as well as the quality of the products using measurement and testing techniques. State how the client will do acceptance.

Risk Management

Summarize the risks you foresee. Show how you will manage risks by having a Risk Management Plan that identifies and evaluates risks and explains how you will react to and control risks.

Project Documentation and Communication

List here any applicable documents: standards, regulations, plans provided by outside contractors, and so forth. Provide an outline of the major re-

Table 5.1: Communications Summary Table

Who	*What*	*When*	*Where/How*	*Why*
Client	Costs	Phase 1	Meeting	Approval

ports that will be produced, such as the status reports, milestone reports, and other project documents. Who receives each report, and what is his or her responsibility after receiving it? What is the distribution schedule?

Explain the two classes of documents in the project: user and project management. Outline what documents will be produced, when, and the responsibilities involved.

Consider a communications summary table in a format like the one shown in Table 5.1.

Assumptions

Legal items, ownership, liability, and especially contractual conditions are emphasized here. What if the estimates were based on major assumptions for some risk items? What will you do if the assumptions do not prove to be true? Cover your derriere by insisting on having the right to renegotiate the contract.

Summary

Repeat the three or four most important messages in the Project Charter, normally related to the project aim and objectives, cost, time, and the resources involved.

List of References

Provide details of key reference documents, including approval and decision documents, relevant publications or communications, and theoretical or academic material.

Approval

Leave space for approvals by the Development Manager, the client sponsor, and all of the crucial resource providers.

Appendixes

Include as appendixes the details of the plan. For example, the WBS, financial tables, risk tables, and design drawings may be included. Do not forget to reference these appendixes in the main part of the document.

Project Plan: Example

KLSJ Consulting

October 14, 2002

Project Proposal

Ottawa–Carleton

Water Park

14 Palsen St., Ottawa, ON, Canada, K2G 2V8

Executive Summary

The following document provides the *Project Plan* for the Ottawa–Carleton Water Park being developed for Carlington Aquatic Parks. The report describes the background and objectives for the project and explains the strategy behind the project management structure proposed by KLSJ Consulting. To meet the proposed targets, *approval of this Project Plan by the owners of Carlington Aquatic Parks is required* within 30 calendar days *from the date of this report.*

The management philosophy for the project is that KLSJ Consulting will manage the design, development, and construction of the park and then hand over to the operations management team prior to the official opening. Carlington Aquatic Parks will retain approval authority for all critical design, marketing, and financial decisions throughout the project. Project team resources include the Project Manager, Project Leader (Design and Construction), Team Leader (Legal), Team Leader (Financial), and Risk Manager. The design and construction team also includes three junior members for the construction period. Marketing will be carried out by a contracted firm overseen by the Project Manager.

The Work Breakdown Structure (WBS), included as Appendix A, describes the activities and schedule for the project. It is separated into three subordinate levels, each representing increasing levels of detail: phases, then functions, and then tasks. In some cases, project tasks are still rather lengthy; therefore a subordinate WBS with increased granularity will be developed for each functional area. The schedule also identifies the "critical path" of activities and milestones, which must be met to achieve the target opening date of May 21, 2005.

Project duration is from September 2002 to May 2005, a period of 32 months. The total project cost is Can$12.45 million (US$9.1 million), based on a Class B estimate, with a range of +25% to −10%. The project risk is high for Phases 1 (Concept) and 2 (Planning), but once the issues of land approval and capital investment have been addressed, the project risk becomes low for Phase 3 (Execution) and beyond. Risk management and mitigation are dealt with at length in the document, along with alternatives for balancing the requirements of time, cost, and quality.

Contents

Appendixes

Background

Dan Milks, president and CEO of Carlington Aquatic Parks of Ottawa, has identified the *market need and potential benefits* of a water park in the National Capital Region. Mr. Milks envisions a family-oriented water and amusement facility that would provide a recreational outlet for visitors and local residents in a city lacking this type of outdoor recreational attraction. Mr. Milks issued a *Project Concept* document in September 2002 and began the process of marketing his idea to the local government and investors. The concept is based on a Spring 2005 opening for a water recreation facility located within the region, costing about Can$12 million (US$9 million) with a further $4 million available for future development.

The size and cost of the project guarantee that there will be a large number of organizations and individuals directly involved or indirectly interested in the development of the park. A list of key stakeholders includes the following:

- Owners
- Investors
- Government (provincial, regional, and municipal, including the City of Nepean, the Region, and the Ontario Municipal Board)
- Community groups
- Contractors
- Employees
- Related business partners

Given the substantial support throughout the region for the idea,[1] Carlington Aquatic Parks has now decided to proceed with the development of the Water Park project. KLSJ Consulting has been contracted to develop the *Project Plan* in order to map out the process for the successful and timely completion of the Ottawa-Carleton Water Park.

Objectives

The overall *objective* of this Project Plan is to describe how KLSJ Consulting will successfully complete the design and construction of the Ottawa–Carleton Water Park facility on behalf of Carlington Aquatic Parks. The Water Park will be completed and ready for a scheduled opening on May 21, 2005, at a total cost of Can$12.450,000. Note that this cost is a Class B estimate with a range of +25% to −10%.

The intention of the owners (Carlington Aquatic) is to create a top-quality recreational facility with the following criteria:

- Capacity of 4500 guests per day, growing to 7000
- Unique attractions, including an outdoor wave pool, an artificial river, and water slides

- A proposed site in the western portion of Ottawa–Carleton's greenbelt that is within a 45-minute drive for over 1 million residents of eastern Ontario and western Quebec
- Parking area adjacent to the Water Park capable of accommodating 1000 cars
- Serviced by local roads, bus transportation links, and bicycle trails

Project management theory[2] speaks of a *triple constraint:* time, cost, and quality. Given the fixed target date for the opening of the park and the low probability that extra funding could be located on short notice, any project related problems are likely to result in a decrease in the number of attractions or their quality. KLSJ Consulting can provide and execute an effective project management system that will ensure that the intended cost, time, and quality are delivered.

The major *deliverables* from the project include the following:

- Feasibility Study
- Business Plan
- Project Plan and Project Budget
- Risk Management Plan
- Environmental Assessment
- Zoning Approval
- Architectural and Engineering Design
- Project Presentations (for political and community groups)
- Investor Commitments
- Marketing Plan
- Operations Staff Guidance (financial, marketing, operations)
- Acceptance Plan
- Project Closing Final Report

Project Team

The Project Team will consist of five core members for the duration of the project, plus three additional persons required to coordinate activities during the construction portion (Phase 3, Execution). The basic organization of the team is shown in Figure 5.2. Close horizontal communications among the team members will be essential.

The specific *responsibilities* of the team members are as follows:

- *Project Manager (Karen Dhanraj).* Overall responsibility for all aspects of the project on behalf of the CEO and owner group of Carlington Aquatic Parks. The Project Manager has primary responsibility for the following:

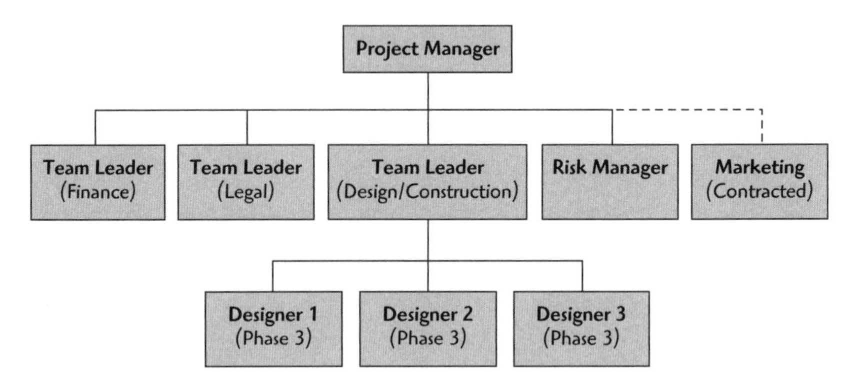

Figure 5.2 Project Team.

- Coordination and communication with outside agencies
- Interfacing with community groups, the Capital District, and municipal, regional, and provincial governments
- Providing direction and guidance to other team members
- Maintaining the overall project plan (schedule, cost, and resources)
- Coordinating and producing all project documentation
- Coordination and oversight of the marketing agency (outsourced)
- Membership in the Risk Management Working Group

▲ *Team Leader (Design and Construction) (Scott Kennedy).* Responsible to the Project Manager for coordination of all activities related to design, engineering studies, permits, and construction. The major goal of the Project Leader is to produce a high-quality water park. Specific responsibilities include the following:

- Supervision of the preliminary design and scale model as well as a detailed final design
- Coordination of the physical construction of the water park and related facilities
- Oversight of the various contractors involved in the construction phase
- Overall quality control of the water park
- Change control with respect to design and construction
- Supervision of the three additional project team members during Phase 3 (Execution)
- Membership in the Risk Management Working Group

▲ *Team Leader (Legal) (Laverne Fleck).* Responsible to the Project Manager for legal counsel related to project activities as well as contract management and interface with regulatory authorities on legal issues. Specific responsibilities include the following:

- Drafting, review, and management of all contracts with contractors (contract management)

- Negotiation of the long-term site lease
- Membership in the Risk Management Working Group

▲ *Team Leader (Finance) (Steve Jackson)*. Responsible to the Project Manager for financial issues related to the project. Specific responsibilities include the following:

- Development and monitoring of business plans, budgets, and project cash flows
- Lead responsibility for locating and confirming investors in conjunction with the President of Carlington Aquatic Parks, the ownership group, and banks and investment firms
- Payment of all project expenses incurred by the project staff
- All project information technology (IT) systems, both for internal project use and for ongoing use by the Water Park operations
- Membership in the Risk Management Working Group

▲ *Risk Manager (Jim Harris)*. Responsible to the Project Manager for coordinating the Risk Management Program. Specific responsibilities include the following:

- Advising on risk management issues
- Ensuring that risk management documentation is current
- Chairing the Risk Management Working Group

▲ *Team Members (Design and Construction)*. Responsible to the Team Leader (Design and Construction) for overseeing specific aspects of the Water Park construction program. Workload analysis indicates that three persons will be required, with each responsible for one of the following areas:

- Procurement, installation and related construction of attractions (slides, pools, etc.)
- Construction and renovation of on-site facilities (buildings, lighting, landscaping)
- Construction of off-site facilities (parking lot, connecting roads)

Project Management Strategy

The overall project strategy calls for the initial planning and preparation to be completed by the end of Phase 2, before construction begins in Phase 3. Specifically, land approval, investor commitments, and conceptual design all will be completed before formal construction is undertaken.

The intention of Carlington Aquatic Parks is that all responsibility for the management of the development and construction activities will be contracted to KLSJ Consulting. Carlington will remain actively involved in the marketing of the project to political offices, community groups, potential investors, and future customers. Carlington also will retain approval authority for all strategic aspects of the project unless that is specifically delegated to KLSJ. Thus, Carlington is expected to approve financial plans and major expenditure (above Can$10,000 or US$7,500), investors, architectural and engineering designs, and marketing activities.

KLSJ will act as the overall project manager while contracting specific aspects of project execution to technical firms selected by KLSJ and approved by Carlington. Functions to be contracted to outside experts include the following:

- Environmental assessment (including environmental, architectural, and traffic studies)
- Engineering site services study
- Engineering design and construction
- Marketing

Also, KLSJ will locate candidate firms for ongoing operations management of the facility.

KLSJ works exclusively with Microsoft software, including Project for project management, Excel for financial reports, Word for reports, and PowerPoint for presentations. All contractors will be expected to integrate with these systems. The specific IT operating systems for ongoing facilities management will be determined during the development of the project.

Scope of Activities

The WBS for the Ottawa Carleton Water Park is contained in a Microsoft Project file and is shown in Appendix A. The plan contains 153 activities that cover a period of nearly 32 months. The activities are subdivided into three levels: phase, function, and specific tasks. Grouping activities by phase allows the Project Manger to get an overview of the interrelationships of different activities as well as the overall flow of the schedule. Grouping activities by function allows the Project Manager and other team members to track the interaction and sequencing of activities within a specific functional area.

The project is initially subdivided into five phases:

- *Phase 1: Concept.* This phase involves developing and assessing the feasibility of the overall project concept. Key activities include the owners' partnership agreement and the initial feasibility study.
- *Phase 2: Planning.* This phase involves the development of plans and strategies required in managing the project. Key activities include creating the business plan and project plan, completing preconstruction (engineering) studies, and locating investors.
- *Phase 3: Execution.* This phase involves the actual construction of the facility and also includes an increased focus on marketing.

- *Phase 4: Handover to Operations.* This phase involves the process of finalizing the facility, preparing it to open, and turning over operational control to the park manager.
- *Phase 5: Closing.* This phase involves internal project management functions required to close the project, including the project review and the final report to the client.

The project also was subdivided into six functional areas, each one the responsibility of a project team member, as shown in Table 5.2.

To keep the WBS activity list under 200 items, as specified by the project sponsor, many of the individual tasks are of much longer duration than the standard three to five days. Therefore, the *Master WBS* will be augmented by a series of six subordinate WBS, one for each of the six major functional areas, to define the specific activities involved in each of these tasks more clearly.

A formal scope control process will be put in place to manage changes in scope as part of the overall change control system. Change cutoff dates are scheduled in each phase to allow changes to be incorporated early enough to minimize disruption to other WBS activities. A formal review and report to the client at the end of each phase provides a second opportunity to review changes to the scope of the project.

A scope management plan integrated with other control processes (schedule, cost and quality) will be promulgated. This plan will have the following general characteristics:

- Change requests must be written and fully substantiated.
- Each change request must be assessed as to its impact on cost and schedule.
- A tracking system will monitor changes.
- Approval levels will be used for authorizing changes.

Table 5.2: Functional Areas and Project Team Leaders

Function	*Lead*	*Majority of Effort*
Project management	Project Manager	Phases 1 to 5
Contract management	Team Leader (Legal)	Phases 2 and 3
Financing	Team Leader (Finance)	Phases 2 and 3
Political and legal	Team Leader (Legal)	Phases 1 and 2
Construction	Team Leader (Design and Construction)	Phases 2 and 3
Marketing	Project Manager (outside contract)	Phases 3 and 4

Project Schedule

The project is scheduled to start on September 16, 2002, with the grand opening of the Water Park expected on Saturday, May 21, 2005 (Victoria Day weekend). Initially, the schedule was developed forward from a set start date. This resulted in an opening date in February 2005 with an unrealistic winter construction period. The schedule then was worked backward from the 2005 Victoria Day weekend to arrive at a workable project. For a more detailed depiction of the schedule, see the Work Breakdown Structure in Appendix A.

This WBS is divided into five phases. Phase 1 mainly involves determining the feasibility of the project. Once the viability of the Water Park has been determined, the project can proceed to Phase 2, in which zoning approval and investor capital are sought. With these attained, Phase 3, which includes construction activities, can commence. Successful completion of construction permits the handover of the Water Park to the operations team in Phase 4. Once operations are in place, KLSJ can close the project in Phase 5.

Key milestones from the critical path within each of the five phases are shown in Table 5.3.

Table 5.3: Critical Milestones

Milestone	*Date*	*Key Stakeholder*	*Key Trigger Event(s)*
Phase 1: Concept			
Approval of project charter/partnership	November 12, 2002	Project Manager	Creation of core project team
Selection of preliminary Water Park site	January 21, 2003	Project Manager	Assessment of site options
Approval of Phase 1 and end of phase review	March 17, 2003	Project Manager	Presentation to Capital District authorities
Phase 2: Planning			
Approval of project plan	April 28, 2003	Project Manager	Approval of Phase 1 (Concept)
Approval of environmental assessment	February 2, 2004	Project Manager	Completion of environmental study

(Continued)

Table 5.3: Critical Milestones (*Continued*)

Milestone	*Date*	*Key Stakeholder*	*Key Trigger Event(s)*
Zoning approval from City of Nepean	February 2, 2004	Project Manager	Conduct environmental assessment, Capital District site approval
Approval of Phase 2 and end of phase review	February 2, 2004	Project Manager	Zoning approval, confirmation of investors
Phase 3: Execution			
Approval of construction permit	July 20, 2004	Project Leader	Completion of detailed engineering design
Approval of construction	November 16, 2004	Project Leader	Completion of construction, installation of attractions, landscaping
Approval of Phase 3 and end of phase review	November 30, 2004	Project Manager	Approval of construction
Phase 4: Handover to Operations			
Trial opening date	April 26, 2005	Project Manager	Final site cleanup and set-to-work preparations
Handover to the operating team	May 10, 2005	Project Manager	Post-trial modifications
Grand opening ceremony	May 18, 2005	Operations Team	Handover to operating team
Phase 5: Closing			
Completed postproject review	June 7, 2005	Project Manager	Handover to operating team
Project completion	June 14, 2005	Project Manager	Project presentation to client, financial reconciliation

Table 5.4: Noncritical Milestones

Milestone	*Date*	*Key Stakeholder*	*Key Trigger Event(s)*
Phase 1: Concept			
Signing the partnership agreement	November 11, 2002	Team Leader (Legal)	Create core project team
Approval of the architectural design	June 23, 2003	Project Leader	Approve Phase 1 (Concept)
Phase 2: Planning			
Investors' commitment	December 8, 2003	Team Leader (Finance)	Approve detailed business plan
Site approval from the Capital District	December 22, 2003	Project Manager	Presentation to Capital District

In addition, there are other *important milestones* that are not part of the critical path, yet are essential for schedule progress, as shown in Table 5.4.

Several features within the WBS allow for tight schedule control. There is a specific deadline for changes in each phase. Hence, if changes are requested, it is possible to allow them without disrupting major deadlines. To provide further control, each phase concludes with an end-of-phase review and approval milestone. In addition to these built-in schedule controls, KLSJ will use the following:

- A schedule change management plan that details how changes to the schedule will be managed
- Performance reports before the end of each phase, providing information about which planned dates have been met and which have not
- Change requests, required before the deadline for changes, that will alert KLSJ to schedule delays or accelerations

Project Costs and Financial Analysis

Cost Summary

The project is estimated to cost Can$12,448,250, which is a Class B estimate (indicating a possible range of −10% to +25%). This amount includes a 10% profit margin for KLSJ Consulting, as agreed on with the customer, as well as applicable taxes. KLSJ has confidence in this estimate because the majority of the figures used are based on catalogue prices for equipment and similar construction work undertaken at other

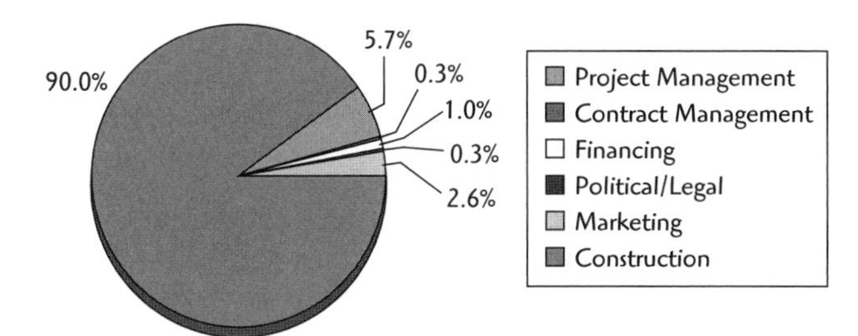

Figure 5.3 Total cost by project function.

water parks. However, some of the larger cost estimates (for example, excavation and landscaping) are based on rule-of-thumb calculations for the magnitude of work specified. Where applicable, a contingency of 10% has been built into the cost estimate to account for anticipated variation in prices.

Costs By Function and Phase

Figures 5.3 and 5.4 depict the costs by project function and by phase. They highlight the fact that the vast majority of project costs will occur in Phase 3 during the actual physical construction of the Water Park, as one would expect. This also emphasizes the fact that the project should not proceed beyond Phase 2 until 85% of the funding has been secured. All project activity will cease at the end of Phase 2 until investors are formally committed, as discussed further under "Risk Management." Detailed breakdowns of the project costs by project phase, project category, and cost category (chart of accounts) are provided in Appendix B. Two items of interest to the client are as follows:

- Project management and contract management account for roughly 6% (Can$683,350) of the total cost, not including the profit margin.
- Only 4.5% (Can$511,550) of the total cost is expended during Phases 1 and 2, meaning that there is relatively little financial risk exposure until the project is committed to begin Phase 3.

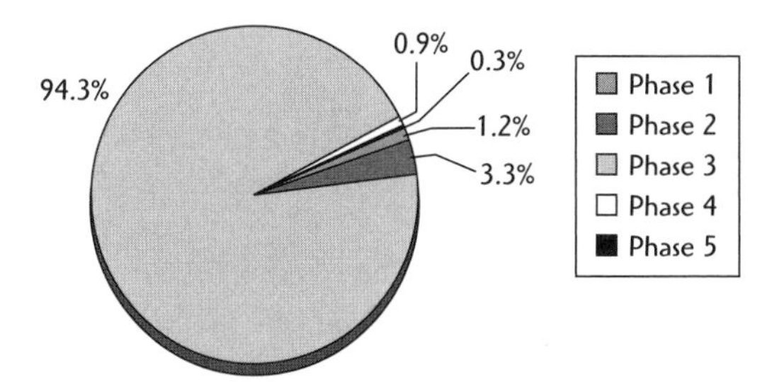

Figure 5.4 Total cost by project phase.

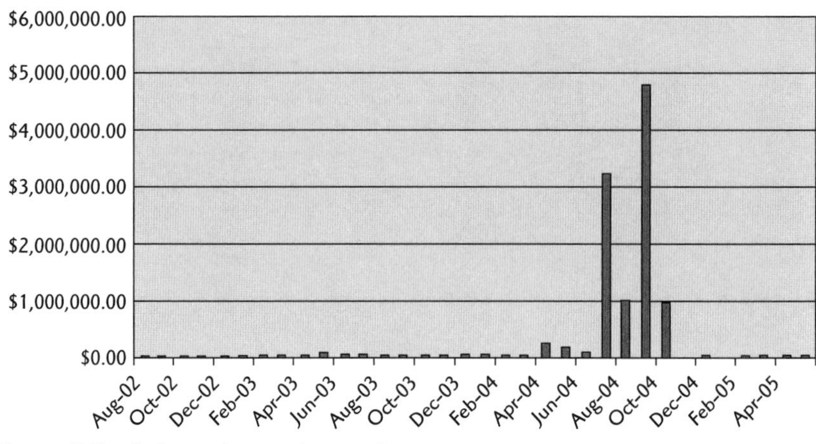

Figure 5.5 Cash requirement by month.

Cash Flow and Cash Management

Figures 5.5 and 5.6 depict the monthly and cumulative cash requirement for the project, based on current cost estimates and normal contracting payment policies. When activities were more than one month in duration, it was assumed that roughly equal payments would be made at each month end based on specified contract performance. The charts again point out the imperative to secure all funding prior to the commencement of Phase 3. This funding will have to be liquid and readily available to KLSJ at all times to satisfy contractor payments. Failure to meet the scheduled payments with contractors may result in work stoppage or even legal action against the client. For instance, during the four months between July and October 2004, the project will consume approximately Can$2 million per month, the majority of which will be for construction and materials. Details of access to bank accounts and other sources of funding will be the subject of separate discussions.

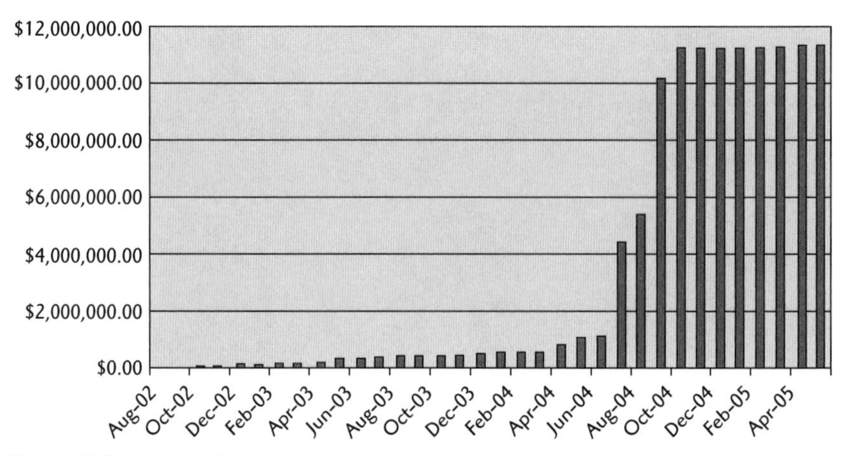

Figure 5.6 Cumulative cash requirement.

Cost Control Strategy

Other than the specific risk management strategies outlined separately, KLSJ will implement specific actions designed to minimize the possibility of cost overruns. These actions will include but are not limited to the following:

- Fixed-price contracts with contractors
- Substantial penalty clauses for nonperformance of any material aspect of a contract
- Where prudent, insurance coverage against specific perils (third-party liability, acts of God, nonperformance of high-risk contracts)
- When necessary and in consultation with the client, deliberate trade-offs between time, quality, and cost to remain within the budget

Resource Estimates and Leveling

The WBS list of activities was used to estimate the personnel resource requirements needed to manage the project. Each activity was assigned to the appropriate member(s) of the project team with an estimate of the level of effort required for each activity. In several cases, noncritical activities were delayed until less work-intensive periods to avoid overtasking a particular member of the project team. However, in no case was it necessary to extend or delay a critical path activity because of a lack of project management resources. It thus was confirmed that the core project team will be sufficient to carry out the estimated work with the exception of the construction during Phase 3 (Execution). An additional three persons will be required during this period to assist the Project Leader in overseeing the actual construction of the Water Park.

Project management resources have been costed at standard company chargeout rates, which include overhead and profit. These costs have been incorporated into the total project cost and are captured separately in cost account C011 (see Appendix B). Resource costs have been levied only for those periods in which a team member is engaged on project business. Finance and Legal members, for example, are not required full-time on the project.

Quality Control and Performance Measures

The project team will employ a combination of qualitative and quantitative methods to monitor the project. Quality control measures ensure that the project is able to meet the specifications intended for the facility. The following quality control measures will be implemented to ensure that the project meets all customer needs and specifications. They will be performed by the project management team, the contractors, or independent outside agencies, as appropriate:

- *Control of risk* via the Risk Management Working Group
- *Control of suppliers, contractors, and subcontractors* through a "qualified seller" listing
- *Control of work performed* through inspection, procurement audits, destructive testing (concrete), and nondestructive testing (attractions)
- *Control of design and engineering* through strict document control, document change control, formal review processes, and independent consultation where necessary
- *Control of financial transactions* through segregation of responsibilities, expenditure controls and limits, limited payment authority, and an independent audit
- Where applicable, the *use of statistical tools* such as process capability and statistical process control to ensure that equipment and processes operate within specified tolerances

Risk Assessment

Recognizing and managing potential risks before they impact the project is an essential aspect of project management. Risk management is a proactive and iterative process that is vital to controlling costs, meeting deadlines, and producing quality results.

KLSJ Consulting has conducted a thorough risk assessment of the Water Park project. The criteria for evaluating risk are presented in Appendix C, along with a summary of the key risks that have been identified. Complete details of the risk assessment are outlined in the Risk Management Plan.

Based on the risk assessment and the high level of risk in the early phases, KLSJ assesses the Ottawa–Carleton Water Park project as a high-risk project. The two most significant risks that the project faces are:

- Failure to secure funding in time to meet project deadlines
- Political action group(s) successfully petitioning the Zoning Authority, the Capital District, Region, or the City of Nepean, resulting in delays or denial of site approval and subsequent delays in the start of Phase 3 (Execution) to the extent that completion by the target date becomes improbable

Each of these risks has a high probability of occurring and would have a large impact on the project in terms of both time delays and additional costs (see Appendix C, Table 5.6).

Without a commitment from investors, the project will come to a halt at the end of Phase 2. The bulk of the Can$12,450,000 must be guaranteed before Phase 3 for the project team to hire contractors and purchase slides and other equipment. Phase 3 construction alone will require $10,500,000 and should not begin until these funds are available.

Similarly, zoning approval from the City of Nepean is essential for the commencement of Phase 3. If zoning is not approved, not only will Carlington Aquatic Parks need to find a new location but the preliminary Water Park design and scale model may have to be redone completely. A change in location in turn may result in the withdrawal of potential investors and further battles with political groups that oppose an alternative site. In addition, the project team would need to reassess the financial requirements relative to the new site, which may render the current cost estimate invalid.

In an effort to minimize the impact of these two high-risk items, the project has been planned so that only a minimal financial commitment is required until these hurdles are overcome in Phase 2. The estimated cost to get to this point of the project is $513,550 (4.5% of the project cost).

Based on these figures, the initial expenditure of approximately $515,000 is a high-risk investment. However, once financing has been secured and political approval has been received for the land and zoning, the overall level of risk for the Water Park drops significantly (see Appendix C, Table 5.9). At the start of Phase 3 (Execution) the remaining $11,500,000 is considered a low-risk investment.

A Risk Management Working Group (RMWG) has been established and will be responsible for all risk-monitoring and mitigation activities. The group will meet monthly at a minimum to review the risk forms and react appropriately. All changes will be published. Special meetings will be arranged to address emergency items as required.

Although the initial risk management process has identified many potential risks associated with the project, new risks will continue to surface during the two and a half years of the project. The risk management team will monitor the probability and impact of risks constantly and revise the project completion date, costs, and deliverable dates accordingly. Carlington Aquatic Parks will be consulted and advised on such changes if and when they occur.

Project Documentation and Communication

The project staff and contractors will require several management documents during the project. All project documents will be made available to the client and the relevant operations staff at specified milestones or at the end of the project.

Other plans and documents in support of the overall project plan include the following:

- Feasibility study
- Environmental assessment (including the environmental, archaeological, and traffic studies)

- Engineering site services study
- Engineering design
- Business plan
- Financial management plan
- Risk management plan
- Marketing plan
- Operating plan (ongoing operations)
- Complete functional Work Breakdown Structure with the following sections:
    - Project management
    - Contract management
    - Political and legal
    - Financial
    - Construction
    - Marketing
- Partnership agreements
- Interim (end-of-phase) and final (end-of-project) client reports and presentations
- Procurement documents and contracts
- Warranties, clear title, deeds, licenses, registrations, and lease agreements

Assumptions

The project plan described in this report is subject to a vast number of external and internal influences, many of which cannot be projected accurately or defined at the outset of the project. The project plan is founded on the following key assumptions:

- Economic and political environmental factors will continue to support the project concept.
- Community support for construction and ongoing patronage will continue.
- Sufficient debt and equity investors can be located no later than mid-February 2004.
- The budgeted amount of Can$12,450,000 will be sufficient to complete the project as intended.
- The project start-up in September 2002 and opening in May 2005 are fixed milestones.
- A suitable location can be identified and acquired in a timely fashion.
- The environmental assessment will be favorable or mitigation can be done with reasonable effort.

- The investors will be satisfied with the engineering design.
- Contractors of sufficient size and expertise are available.
- Delivery of all equipment is within the time allotted.
- All permits and permissions are secured within the time allotted.
- The weather during Phases 3 and 4 remains within normal seasonal parameters.
- A suitable operations manager can be found at a reasonable salary.
- The trial run reveals no major problems.

KLSJ believes that these assumptions are reasonable. If any of the assumptions prove otherwise, there could be an impact on project cost, time, and/or quality. A risk analysis, as described previously, has been completed, and mitigation action as applicable will be undertaken to minimize these impacts.

Summary

KLSJ Consulting has developed a comprehensive project plan that will allow it to oversee the conceptual development, design, approval, construction, testing, and handover of the new Ottawa–Carleton Water Park. The project plan will maximize the probability of completing the facility as designed, on time, and on budget.

The first step on the road to the completion of this exciting new facility is the approval of the enclosed Project Plan by the managers of Carlington Aquatic Parks. Approval is required within the next 30 days for KLSJ to maintain the Project Plan schedule.

References

1. "Few Waves Yet to Ride," *Link Magazine*, Spring/Summer 1999, p. 26.
2. Rakos, J., *Software Project Management for Small to Medium Sized Projects*, Prentice Hall, Englewood Cliffs, NJ, 1990, p. 152.

Approvals

This plan is approved by:

______________	______	______________	______
Signature	Date	Signature	Date
Karen Dhanraj, Project Manager		Dan Milks, President and CEO of Carlington Aquatic Parks of Ottawa	

Appendix A: Work Breakdown Structure of Ottawa–Carleton Water Park Project

The Work Breakdown Structure of the project is shown in Figure 5.7.

ID	WBS	Task Name	2002–2005 (Q2 2002 – Q3 2005)
1	**1**	**OC Water Park**	
2	**1.1**	**Phase 1 - Concept**	
3	**1.1.1**	**Project Management**	
4	1.1.1.1	develop initial project concept	PM
5	1.1.1.2	create core project team (Design/Fin/Legal)	PM
6	1.1.1.3	*approval of project charter (partnership)*	09/27
7	1.1.1.4	complete feasibility study	PM,TL(Fin)[50%],PL (Design)[50%]
8	1.1.1.5	*approve project concept*	11/22
9	1.1.1.6	assess site options	PM[80%],PL (Design)
10	1.1.1.7	*select preliminary site*	12/06
11	1.1.1.8	develop initial business case	TL(Fin),PM[50%]
12	1.1.1.9	*deadline for Phase 1 changes*	01/03
13	1.1.1.10	*approve Phase 1 (Concept) end phase review*	01/31
14	**1.1.2**	**Contract Management**	
15	1.1.2.1	develop partnership agreement	TL(Legal)
16	1.1.2.2	*sign partnership agreement*	02/07
17	**1.1.3**	**Financing**	
18	1.1.3.1	identify potential investors (equity and bank)	TL(Fin)[50%]
19	**1.1.4**	**Political/Legal**	
20	1.1.4.1	verify contract bidding system	TL(Legal)
21	1.1.4.2	present to NCC authorities	PM[50%],PL (Design)[50%]
22	**1.1.5**	**Construction**	
23	1.1.5.1	develop preliminary design	PL (Design)
24	1.1.5.2	create scale model	PL (Design)
25	**1.1.6**	**Marketing**	
26	1.1.6.1	conduct market research (contracted)	PM[10%]
27	**1.2**	**Phase 2 - Planning**	
28	**1.2.1**	**Project Management**	
29	1.2.1.1	develop project plan (including WBS)	PL (Design)
30	1.2.1.2	*approve project plan*	03/21
31	1.2.1.3	develop project budget	PM[50%],TL(Fin)[50%]
32	1.2.1.4	conduct visits of other waterparks (Canada/US)	PM
33	1.2.1.5	develop detailed business plan	PM[50%],TL(Fin)[50%]
34	1.2.1.6	develop quality assurance system	PL (Design)[40%]
35	1.2.1.7	develop admin procedures	PM[50%]
36	1.2.1.8	conduct environmental study (contracted)	PM[20%]
37	1.2.1.9	*interim environmental report (spring/summer)*	09/22
38	1.2.1.10	*final environmental report*	12/12
39	1.2.1.11	conduct archeological study (contracted)	PL (Design)[20%]
40	1.2.1.12	conduct traffic study (contracted)	PL (Design)[20%]
41	1.2.1.13	conduct engineering site services study (contracted)	PL (Design)[20%]
42	1.2.1.14	*approve site services study*	07/11
43	1.2.1.15	conduct environmental assessment (contracted)	PM[20%]
44	1.2.1.16	*approve environmental assessment*	01/09
45	1.2.1.17	*deadline for Phase 2 changes*	05/09
46	1.2.1.18	*approve Phase 2 (Planning) end phase review*	01/09

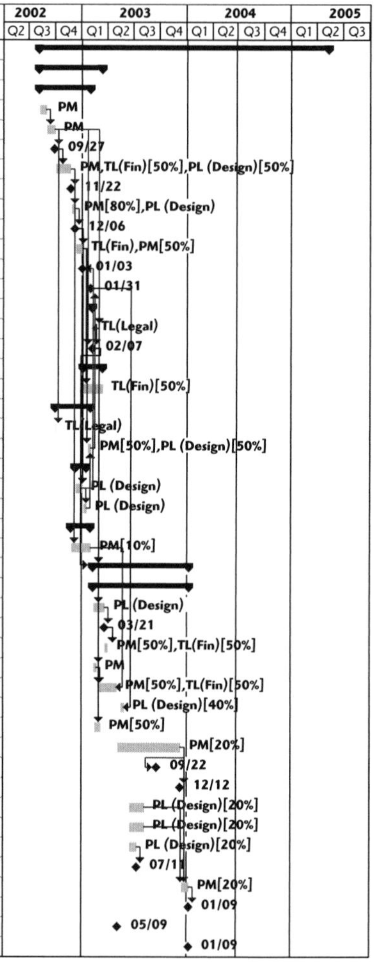

Figure 5.7 Work Breakdown Structure of Ottawa–Carleton Water Park Project.

ID	WBS	Task Name	2002 Q2–Q4 / 2003 Q1–Q4 / 2004 Q1–Q4 / 2005 Q1–Q3
47	**1.2.2**	**Contract Management**	
48	1.2.2.1	select environmental study consultant	PM[20%]
49	1.2.2.2	select environmental assessor	PM[20%]
50	1.2.2.3	select archeological assessor	PL (Design)[20%]
51	1.2.2.4	select traffic engineer	PL (Design)[20%]
52	1.2.2.5	select site service engineers	PL (Design)[20%]
53	1.2.2.6	select marketing company	PM[20%]
54	**1.2.3**	**Financing**	
55	1.2.3.1	*approve detailed business plan and budgets*	05/02
56	1.2.3.2	confirm investors	TL(Fin)
57	1.2.3.3	*investors committed*	10/31
58	**1.2.4**	**Political/Legal**	
59	1.2.4.1	presentation to RMOC	PM[20%],PL (Design)[20%]
60	1.2.4.2	presentation to community groups	PM[20%]
61	1.2.4.3	presentation to MVCA	PM[20%]
62	1.2.4.4	presentation to NCC	PM[50%],PL (Design)[30%]
63	1.2.4.5	*site approval from NCC*	11/28
64	1.2.4.6	negotiate NCC lease	TL(Legal)[50%]
65	1.2.4.7	*NCC lease approval*	12/26
66	1.2.4.8	presentation to City of Nepean	PM[50%],PL (Design)[30%]
67	1.2.4.9	*zoning approval from City of Nepean*	01/09
68	**1.2.5**	**Construction**	
69	1.2.5.1	develop detailed architectural design	PL (Design)
70	1.2.5.2	*approve architectural detailed design*	05/09
71	**1.2.6**	**Marketing**	
72	1.2.6.1	develop marketing plan (contracted)	PM[20%]
73	1.2.6.2	*approve marketing plan*	06/20
74	1.2.6.3	develop brand name/trademark (contracted)	PM[10%]
75	**1.3**	**Phase 3 - Execution**	
76	**1.3.1**	**Project Management**	
77	1.3.1.1	coordinate telephone service	PL (Design)[20%]
78	1.3.1.2	*deadline for Phase 3 changes*	07/30
79	1.3.1.3	coordinate traffic light/road upgrades	TM3(Design)[20
80	1.3.1.4	coordinate bus access	TM3(Design)[2
81	1.3.1.5	coordinate emergency services (fire/medical)	TM3(Design)[2
82	1.3.1.6	develop hotel shuttle service	TM3(Design)[2
83	1.3.1.7	*approve Phase 3 (Execution) end phase review*	10/22
84	**1.3.2**	**Contract Management**	
85	1.3.2.1	select parking contractor	PM[10%]
86	1.3.2.2	develop operations plan	PM[50%]
87	1.3.2.3	hire operations management team	PM[50%]
88	1.3.2.4	select construction management firm	PL (Design)[20%]
89	1.3.2.5	select design engineers	PL (Design)[20%]
90	1.3.2.6	select water/sewage contractor	PL (Design)[20%]
91	1.3.2.7	select road/traffic light contractor	PL (Design)[20%]
92	1.3.2.8	select electrical transformer contractor	PL (Design)[20%]
93	1.3.2.9	*approve operations plan*	06/25
94	**1.3.3**	**Financing**	
95	1.3.3.1	procure site IT operating system	TL(Fin)[30%]
96	1.3.3.2	procure marketing software	TL(Fin)[30%]
97	1.3.3.3	procure HRIS	TL(Fin)[30%]
98	1.3.3.4	develop financial management system for operations (payroll/banking)_	TL(Fin)

Figure 5.7 *Continued.*

ID	WBS	Task Name
99	**1.3.4**	**Construction**
100	1.3.4.1	develop detailed engineering design (contracted)
101	1.3.4.2	obtain construction permits
102	*1.3.4.3*	*construction permit approved*
103	1.3.4.4	procure slides and attractions equipment
104	1.3.4.5	receive slides and equipment
105	1.3.4.6	conduct site electrical
106	1.3.4.7	*site electrical complete*
107	1.3.4.8	conduct water/sewage
108	1.3.4.9	*site water/sewage complete*
109	1.3.4.10	conduct road construction
110	1.3.4.11	*road construction complete*
111	1.3.4.12	conduct excavation/demolition
112	1.3.4.13	renovate existing buildings
113	1.3.4.14	construct new buildings
114	1.3.4.15	construct miscellaneous structures
115	1.3.4.16	construct pools
116	1.3.4.17	install water slides
117	1.3.4.18	install wave pool
118	1.3.4.19	install perimeter fencing and lighting
119	1.3.4.20	install concessions, minor attractions
120	1.3.4.21	install river
121	1.3.4.22	construct parking areas (not paved)
122	1.3.4.23	do landscaping
123	1.3.4.24	*approve construction*
124	**1.3.5**	**Marketing**
125	1.3.5.1	develop advertising and marketing communciations
126	1.3.5.2	*approve advertising and marketing communications*
127	1.3.5.3	conduct initial promotions/advertising
128	1.3.5.4	attract corporate sponsors
129	1.3.5.5	procure tickets and passes
130	1.3.5.6	confirm tourism department promotion
131	**1.4**	**Phase 4 - Turnover to Operations**
132	**1.4.1**	**Project Management**
133	1.4.1.1	prepare for trial opening
134	1.4.1.2	conduct trial opening and evaluation
135	1.4.1.3	*trial opening date*
136	1.4.1.4	conduct handover with park operations team
137	1.4.1.5	*handover to operating team*
138	1.4.1.6	*grand opening ceremony (Park Opening)*
139	**1.4.2**	**Contract Management**
140	1.4.2.1	close out construction contract
141	**1.4.3**	**Construction**
142	1.4.3.1	conduct final site clean up
143	1.4.3.2	carry out modifications/repairs after trial
144	**1.5**	**Phase 5 - Closing**
145	**1.5.1**	**Project Management**
146	1.5.1.1	close out project documentation
147	1.5.1.2	conduct post project review (Lessons Learned)
148	1.5.1.3	*complete post-project review*
149	1.5.1.4	project presentation to client
150	1.5.2	*Project Complete*
151	**1.5.3**	**Financing**
152	1.5.3.1	do final payments and banking
153	1.5.3.2	do financial reconciliation

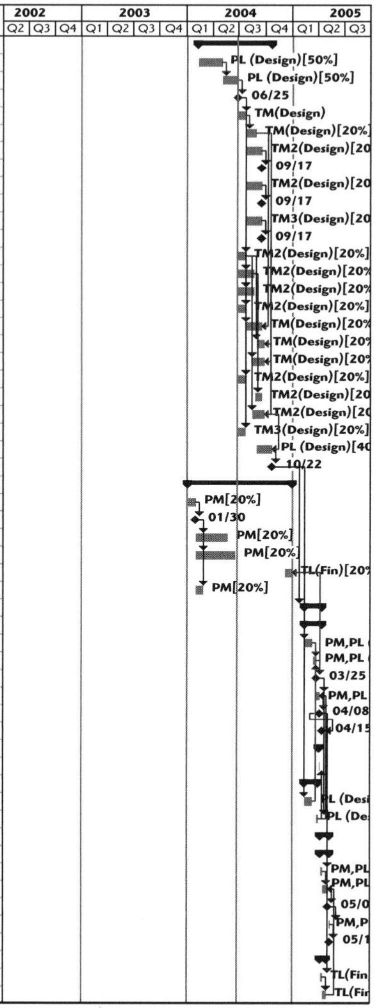

Figure 5.7 *Continued.*

Appendix B: Financial Information: Ottawa–Carleton Water Park Project

The costs for the project are shown in Table 5.5.

Table 5.5: Ottawa–Carleton Water Park Costs ($Can)

Totals by Project

Phase	*Fixed Costs*	*Resource Costs*	*Total Costs*
Phase 1. Concept	$25,000	$109,000	$134,000
Phase 2. Planning	$192,000	$185,550	$377,550
Phase 3. Execution	$10,584,000	$134,100	$10,718,100
Phase 4. Handover to Operations	$70,000	$28,500	$98,500
Phase 5. Closing	$0	$33,000	$33,000
Total	**$10,871,000**	**$490,150**	**$11,361,150**
Profit (10%)	$1,087,100		$1,087,100
Grand Total	**$11,958,100**	**$490,150**	**$12,448,250**

Totals by Project Category

	Fixed Costs	*Resource Costs*	*Total Costs*
Project management	$432,000	$213,050	$645,050
Contract management	$0	$38,300	$38,300
Financing	$34,000	$83,200	$117,200
Political and Legal	$0	$35,400	$35,400
Construction	$10,135,000	$91,100	$10,226,100
Marketing	$270,000	$29,100	$299,100
Total	**$10,871,000**	**$490,150**	**$11,361,15**
Profit (10%)	$1,087,100		$1,087,100
Grand Total	**$11,958,100**	**$490,150**	**$12,448,250**

(Continued)

Table 5.5: Ottawa–Carleton Water Park Costs ($Can) (*Continued*)

Totals by Cost	
Category	
C011. Salaries	$490,150
C022. Site Visits	$5,000
C031. Marketing Services	$45,000
C032. Advertising	$200,000
C034. Promotions	$25,000
C045. Studies	$77,000
C054. Permits	$75,000
C061. Design and Engineering	$365,000
C062. Construction—On-site	$4,695,000
C063. Construction—Attractions	$2,800,000
C064. Construction—Miscellaneous	$550,000
C071. Installed Equipment	$2,000,000
C081. Information Technology	$34,000
Total	**$11,361,150**
Profit (10%)	$1,087,100
Grand Total	**$12,448,250**

Appendix C: Risk Assessment

There is a separate Risk Management Plan document provided. A management overview of that document is shown in Table 5.6.

Table 5.6: Risk Evaluation Criteria

Probability Criteria

Probability Rank	***Description***
High	More than 50% probability of occurring
Medium	Between 25 and 50% probability of occurring
Low	Less than 25% probability of occurring

Impact Criteria

Schedule Impact

Impact Rank	***Description***
High	Delay opening of park beyond July 1, 2005 (more than 6 weeks)*
Medium	Delay opening of park between June 15, 2005, and July 1, 2005 (3 to 6 weeks)
Low	Delay the opening of the park between May 24, 2005, and June 15, 2005 (less than 3 weeks)

Cost Impact

Impact Rank	***Description***
High	Could add more than 20% to cost of project (more than Can$2,500,000)
Medium	Could add 10 to 20% to cost of project (between $1,200,000 and $2,500,000)
Low	Could add less than 10% to cost of project (less than $1,200,000)

*Opening after the July 1, 2005, date is economically unsound, and delays beyond this date probably will result in deferring the opening of the Water Park to the following year.

Table 5.7: Risk Table at Start of Project (Prior to Phase 1)

Probability ***Impact***	***LOW***	***MEDIUM***	***HIGH***
HIGH	7. Construction company has major financial difficulties. 8. Major investor withdraws from project.		1. Failure to secure investors. 2. Political groups delay zoning approval.
MEDIUM		5. Unable to find suitable operations manager.	
LOW	4. Delivery of attractions delayed. 6. Design not approved. 9. Delays in obtaining construction permit. 10. Delays due to inclement weather.	11. Post-trial modifications delay opening.	

Evaluation of Overall Project Risk

The evaluation of overall project risk is shown in Tables 5.7 and 5.8.

Table 5.8: Risk Table at Start of Phase 3

Probability *Impact*	*LOW*	*MEDIUM*	*HIGH*
HIGH	7. Construction company has major financial difficulties. 8. Major investor withdraws from project.		
MEDIUM		5. Unable to find suitable operations manager.	
LOW	4. Delivery of attractions delayed. 6. Design not approved. 7. Delays in obtaining construction permit. 10. Delays due to inclement weather.	11. Post-trial modifications delay opening.	

Risk Management Summary

A summary of risk management is given in Table 5.9.

Table 5.9: Waterpark Risk Management Summary Revised October 8, 2002

Risk	*OPI*	*Probability*	*Impact*	*Status*
1. Funding not secured	TL (Finance)[*]	High	High	Open
2. Political groups delay approval	PM	High	High	Open
3. Unfavorable environmental assessment	PM	Medium	High	Open
4. Delay in delivery of attractions	PL (Design and Construction)	Low	Low	Open
5. Unable to hire suitable operations manager	PM	Medium	Medium	Open
6. Final design not acceptable	PL (Design and Construction)	Low	Low	Open
7. Financial difficulties with construction company	PL (Design and Construction)	Low	High	Open
8. Major investor withdraws	TL (Finance)	Low	High	Open
9. Delays associated with construction permits	PL (Design and Construction)	Low	Low	Open
10. Inclement weather delays construction	PL (Design and Construction)	Low	Low	Open
11. Post-trial modifications exceed three-week window	PL (Design and Construction)	Medium	Low	Open

[*]TL: Team Leader; PM: Project Manager; PL: Project Leader.

Appendix D: Conceptual Design

Preliminary drawings of the Water Park, slides, parking area, roads, and main buildings are available as blueprints at the offices of KLSJ.

6

Communications Plan

Communications Plan: Discussion

A communications plan is intended for the members of the Project Team but also may be given to the client and major stakeholders. It is one of the key project documents but unfortunately is an aspect of project management that often is overlooked. A communications plan is required to ensure the timely and appropriate generation, collection, dissemination, storage, and ultimate disposition of project information.

Before writing a communications plan, the author should have an understanding of the *principles* of communication. A good knowledge of basic communications concepts enables the project team to anticipate potential difficulties or misunderstandings. For example, "filtering" problems are a common source of friction: What is said by the speaker may not be what is understood by the audience. There is also a need to understand and recognize the importance of *non*-verbal communication. For instance, there may be occasions when the importance of gauging a stakeholder's unspoken reactions dictates that a team schedule a face-to-face meeting.

Project *language* is another important factor to consider in developing a communications plan. *Language* in this case means the vernacular. Depending on the nature of the project and the participants, it is likely that a certain amount of technical jargon will creep into conversations and correspondence. There are two important points to consider in choosing the language you will use. First, communicate as much as possible in a language the receiver will understand. This implies that the team may need to use the language required by the client. Second, everyone on the team must make an effort to understand and communicate in a common project language, whatever that may be. For highly technical projects or when the team members have diverse backgrounds, a project glossary can be a useful tool for reducing confusion.

The Project Team must ensure that it communicates effectively with all the stakeholders. Since there typically will be many stakeholders with multiple lines of communication, it is essential to consider each stakeholder (or group) individually to achieve the project's communications objectives. The Project Manager has many choices available for getting the message across: from conversations (in person, by phone, or by videoconferencing), to correspondence (memo, letter, document, or publication), to more technological means (fax, e-mail, voice mail, or even a Website). All this must be planned.

Although communication is an ongoing and continuous process, there is a need for certain regular reports. Things that need to be reported regularly include the project's status and resource usage (monthly or even more frequently). Other reports may be produced when a major milestone is achieved or a phase is completed. Ad hoc reports should be produced whenever issues either arise or are resolved, for example, regulatory requirements, major project risks, or a change in the status of funds. The objective in each case should be to keep the stakeholders apprised of the status of the project and report on any issues that have been or will have to be resolved. Brevity is important: A typical status report should be no more than two to three pages in length and take no more than an hour or two to prepare. It should be clear, succinct, and direct, reporting items either briefly or by exception, depending on the subject area (see "Project Status Report" in Chapter 11).

The issue of meetings always comes up in discussions of communications since so much time is spent preparing for and attending them. There are times when holding a meeting is preferred, such as at the start or end of a phase, at major milestone points, and when a problem crops up that needs several minds to resolve. Many hours can be wasted at meetings if they are not conducted properly, and so it is essential for the Project Manager to know how and when to run a meeting.

Every project needs some sort of archive or library where project communication documents and records can be stored, both paper and electronic. This library essentially represents the project's official "history." For paper storage, the location should be central to (or near) the project, in a reasonably accessible yet secure place. Copies of all official project correspondence should be filed here, with the possible exception of technical manuals, which may have to be kept at the project work site. An Internet site can provide convenient access to all these documents electronically.

Finally, regardless of the size of the project, a stakeholder communications chart (see Table 6.1) is an essential element of the communications plan. Indeed, for small projects, the chart is probably all that is needed. The purpose of the chart is to divide stakeholders into groups, analyze their unique communications needs, and then decide on the most effective means of providing for those needs.

Communications Plan Outline

Executive Summary

This is an optional section that summarizes the major points of the document. It should provide an overview to readers who do not need all the details and help them decide whether they need to read the remainder of the document.

Background

Provide a brief history of the project. State how the project started, why, the original high-level objectives, the deliverables, a ballpark cost estimate, and a summary of the schedule and/or milestones. Some or all of this information may be contained in the Project Concept Document (Chapter 1), and it is perfectly acceptable to repeat the main points verbatim if they are still valid.

The Project Environment

The environment in this case is the communications environment. First, discuss the needs and interests of the stakeholders, especially if they are unique or in conflict (for example, the need for regulatory disclosure may conflict with the owners' desire for privacy). Try to put yourself in the shoes of the audience and examine the communication framework from its perspective.

Next, determine whether your project will be of interest to the general public. Depending on the nature of the project, there may be an opportunity to generate awareness of the project or reach out to local communities for support. Conversely, if there is the potential for negative publicity, it is often better to preempt it with a balanced, informative public relations campaign. These types of issues need to be examined and incorporated into the plan.

Table 6.1: Partly Filled-in Stakeholder Communications Chart

Who	***How***	***What***	***When***	***Where***	***Why***
Client (with the Project Manager (PM))	• Telephone • Informal and formal meetings • Formal documents (status reports and change requests)	• Progress • Problems, changes, or decisions • Acceptance	• Constantly; informally and formally • Monthly status reports • Project closure	• At the site • PM's office • Client's office	• To keep client apprised of changes and delays • To reassure client with plans to deal with the unexpected • To seek approvals

Along the same lines, briefly examine the media within the project environment. Consider the type and quantity of media (e.g., print, radio, television, Internet) as well as their potential interest in the project (positive or negative). Almost every project has news value to some organization, and so it is advisable to be thoroughly prepared to deal with inquiries from the media (or from any stakeholder).

Technology

Other than the environment, technology can be the single most enabling or inhibiting factor in project communications. The tools and techniques available to the Project Team can vary greatly, depending on the location of the project and the geographic distance to key stakeholders. If necessary, the technological limitations in some parts of the world can be overcome, but at a price. Videoconferencing, satellite communications, and wireless computer networks may enable better dialogue, but the cost may be prohibitive for all but the largest projects. In more technologically developed countries a project team has a wide array of communications options at its disposal.

No matter where the project is taking place, all communications technology available to the team should be identified and categorized in terms of relevance. The cost and reliability of various technologies will dictate how specific communications needs can be handled. For example, the availability of e-mail may reduce the need for face-to-face meetings. Conversely, an overriding need for security may exclude unencrypted cellular phones.

The Project Manager may wish to designate that certain technology be used for specific purposes. For instance, it may be decided that all correspondence with external agencies will be typewritten, all meetings with the client will be recorded (publicly), and all weekly reports to headquarters will be via secure voice communication.

Communications Objectives

The communications objectives should be in concert with the overall project objectives, and so it is important to refer to the Project Concept or Proposal and the Project Plan for guidance at this point. One of the more obvious objectives is good communications among the project team as well as with customers and other stakeholders. Other objectives, depending on the nature of the project, could involve generating public awareness, political support, buyer interest in the end product, or even follow-on business for the project team.

Objectives lead to activities, which produce specific, measurable outcomes that can be measured and assessed. Therefore, whatever the objectives, you must phrase them carefully to ensure that they are congruent with the high-level project objectives and performance management framework.

Communications Planning

This is where you describe the methodology to be used for crafting the communications plan. Remember that the document you are writing is the plan itself, and so this section should not repeat material from other parts. Instead, focus on the communications planning process, especially how objectives will be turned into concrete actions.

In describing the communications planning process, be sure to document what steps are involved, who will be consulted, how often the plan will be reviewed, and how feedback will be gathered and used to improve both the plan and the process.

Communications Messages

Projects, especially large ones, attract a considerable amount of attention. Therefore, it is very important to speak with one voice. This is essential for both internal and external communications, since clarity and consistency enhance the professional reputation and productivity of a project team.

Prepared messages should be crafted well before the project is materially under way, so that the team will be able to deal effectively with queries from stakeholders, the media, and the public. The project should have a *positioning statement* that describes what the project is about, why the project is taking place, and what its benefits are (and to whom). If necessary, the statement can be broken down into several supporting messages and even expanded in backgrounders. This is particularly effective for large, complex projects that may have impacts across a broad spectrum of the general public.

There also may be cases where a project involves certain specific risks that will, if they occur, have media or legal implications. In these circumstances, it is best to have a well-thought-out position prior to the event so that the issue can be dealt with effectively and rationally. Issues such as labor disputes, environmental damage, and loss of life often demand immediate and very public responses that should not be left to chance. A comprehensive communications plan will have anticipated these events and created possible media responses that are in keeping with company and/or client policies.

Audiences

Audiences should be divided into internal and external groups and further subdivided into categories with similar interests or needs. This section should not be very long provided that the Stakeholder Communications Chart is used.

Each distinct audience category should be discussed in detail in the chart. It is important to identify only the stakeholder groups that have unique needs; otherwise, the chart will become unnecessarily long and repetitive.

Priorities and Constraints

You should identify the communications priorities and requirements during the completion of the Stakeholder Communications Chart. For instance, regular contact with a key regulatory agency may be critical to receiving approval to proceed with a certain project phase. Investors may need weekly updates to ensure continued support for the project. In either case the chart serves as a straightforward way to scan all the major communications needs and select those which will be project priorities.

It is also important to state why a particular aspect of the project plan is a priority, and this should be done within the context of any threats, risks, or opportunities that have been identified. Areas for special attention include any communications issue that, if neglected, could lead to project failure. For example, failure to keep political figures apprised of the benefits of the project could lead to bureaucratic delays, thus jeopardizing the overall schedule.

Pay attention to any communication constraints in the project, such as the need for privacy or even secrecy, depending on the nature of the project. Outsourcing or subcontracting also can be a constraint since it causes the Project Team to lose control over parts of the communications process. Regardless of the source, all major constraints should be discussed in terms of their impact on the team's ability to communicate effectively and on any other area of the project.

Responsibilities

As is emphasized throughout this book, it is very important to define the responsibilities of key project members. This can be done in narrative format or in the form of an organization chart or a combination of both. Pay particular attention to defining who is authorized to communicate externally, whether it is to the media, investors, regulators, other stakeholders, or the general public. Depending on the size or sensitivity of the project, the Project Manager, one of the senior project members, or even a dedicated communications officer could be given overall responsibility for communications. This includes the production and implementation of the communications plan.

Performance Reporting

One of the primary objectives of project communications is to report on project performance. The purpose of reporting is to inform stakeholders about schedule progress, resource consumption, changes in risks or assumptions, and any other project-related milestones and goals. The status report informs the world about progress (what has been achieved since the last report compared to the baseline plan), forecasts (the predicted cost and completion date), and any problems or issues encountered or resolved. See Chapter 11 for a discussion of status reporting. If desired, the Project Manager can specify the format for each type of report and include templates as Appendixes to the Communications Plan. This is an excellent way to facilitate consistency and completeness in project communications.

Depending on the size and complexity of the project, reports may be prepared regularly or by exception. Reports also may be regular for some stakeholders but ad hoc for others, again based on the specific needs identified in the Stakeholder Communications Chart. The chart should identify who receives each report and what his or her responsibility is after receiving it.

Documentation

As was stated at the start of this chapter, communications is a critical aspect of any project. Inevitably, projects generate a great deal of documentation that must be accessible to the Project Team and other stakeholders. Every project management office should set up a system for creating, reproducing, distributing, tracking, filing, and archiving all written and electronic communications. In addition, projects that use more technology-based means of communicating (voice mail, e-mail, streaming video, videoconferencing, etc.) should consider how best to preserve these types of media for later reference. Paperless environments have many benefits, but they can make it difficult to reconstruct the chain of events that led to key project decisions.

It is important to distinguish between what is purely project management documentation (such as plans) and items that the user or owner will need once the project is complete (such as user guides). In a crunch, project teams may try to skip the production of the project management documents. Do not allow this to happen.

Administrative Closure

As the project wraps up, various activities need to take place. The project team should ensure that there are no loose ends, that all outstanding communications have been resolved, that archives and acceptance documentation are complete, and that any lessons that were learned are captured for future reference. Many of these items need not wait until the closure phase of the project but can be done throughout all the phases.

Project documentation, particularly anything that describes the details of the product (specifications, drawings, test results, etc.) must be complete and correct. Test results are especially important since they are used to verify that the team has performed its task. Records should be cataloged and, as described above, separated into items that are to be turned over to the client and those which are strictly project documents.

The last and most important piece of communication for both the Project Team and the client is the formal acceptance document. Regardless of its form, the results of the acceptance should be distributed to all relevant stakeholders and the original should be kept in the permanent files at the company's headquarters.

A note of caution on archiving: The Project Team should never lose sight of the fact that it, or its successors may be called on to do follow-up work, work on a new but related project, or to defend themselves in civil or criminal court. For all these reasons and many more, it is very important to have a well-documented, cross-referenced set of project records for future retrieval. This archiving helps create a knowledge network that can be leveraged for use in the future. In contrast, project files thrown into boxes without recording of their contents or location as the project office is hastily dismantled and people run off to the next assignment is a potential recipe for disaster.

For more on the specific duties related to project closure, see Part 4.

Communications Plan: Example

KLSJ Consulting

October 15, 2003

Communications Plan

Ottawa–Carleton

Water Park

14 Palsen St., Ottawa, ON, Canada, K2G 2V8

Executive Summary

The Communications Plan that follows provides a detailed description of how communications will be managed and controlled for the Ottawa–Carleton Water Park project. It describes the relevant communications environment, the communications objectives and priorities, the messages that are to be emphasized, and the audiences that are to be considered.

The Team Leader (TL) has direct responsibility for all routine project communications, and the Project Manager retains overall authority for certain aspects of the plan, particularly relations with politicians, investors, and the media. One of the principal responsibilities of the TL is the preparation and delivery of project performance reports. Performance measurement is to be done in accordance with KLSJ's ISO 9001:2000 standard and shall be coordinated closely with the project's Quality Manager.

As with all projects, proper documentation and archiving are an important part of communications management and are to be done in accordance with normal KLSJ procedures. Documentation that must be handed over to the client upon project completion is to be separated and secured off the job site to avoid possible loss or damage.

Owing to the scope of this project, administrative closure will be done through a formal acceptance and handover procedure. Copies of all such documents are to be retained and archived by the Project Team prior to closure.

The Stakeholder Communications Chart (Table 6.2 in Appendix A) provides a summary of all stakeholder groups and how they are to be managed with respect to communications.

Contents

Appendix

Background

Dan Milks, president and CEO of Ottawa–Carleton Aquatic Parks of Ottawa, has identified the *market need and potential benefits* of a water park in the National Capital Region. Mr. Milks envisions a family-oriented water and amusement facility that would provide a recreational outlet for visitors and local residents in a city lacking this type of outdoor recreational attraction. Mr. Milks issued a *Project Concept* document in September 2002 and began the process of marketing his idea to the local government and investors. The concept is based on a spring 2005 opening for a water recreation facility located within the region.

The size and cost of the project guarantee that there will be a large number of organizations and individuals directly involved or indirectly interested in the development of the park. A list of key *stakeholders* includes the following:

- Owners
- Investors

- Government (provincial, regional, and municipal, including the Region of Ottawa–Carleton, and the Province of Ontario Municipal Board)
- Community groups
- Contractors
- Employees
- Related business partners
- Competitors

The Project Environment

Ottawa is the capital of Canada and thus receives a larger than average share of media coverage for a city of its size. All national and several regional media organizations have offices in Ottawa, as do many national and international special-interest groups. Media focus is primarily on politics, but issues such as environmental protection, human rights, architectural and cultural heritage, free trade, and intellectual property protection could have an impact on the project.

As the nation's capital, Ottawa has a unique hierarchy of federal, provincial, regional, and municipal governments, including the National Capital Commission (NCC), which has special jurisdiction over all major construction in the Ottawa area. Citizens of Ottawa are known to be politically active in regional issues, particularly when an issue relates to the development of green space for commercial purposes, which they view as detracting from the local community. Other important external stakeholders include the suburb of Nepean (location of the park) and the Chamber of Commerce, both of which are very supportive of this proposal. The city building inspector and the regional fire marshal, both of whom are known to be strict enforcers of municipal by-laws, also must be considered.

The Ottawa area has six television stations, 18 radio stations, and three local newspapers. It is also in close proximity to Canada's two largest media centers, Toronto and Montreal, which have a combined coverage area that takes in one-quarter of the population of Canada.

Communications Technology

Canada is a technologically sophisticated country, and Ottawa is considered one of the leaders in communication not only in Canada but in the world. All manner of communications technology is available to the Project Team, with the possible exception of secure voice and data at the work site. Regular telephone lines (voice satellite uplink) will be used between the site and the office, but handheld phones and radios also will be employed until the site is fully developed.

Since almost all stakeholders are in the local area, the team can decide which mode of communications to use in various circumstances. As a rule of thumb, all decision documents or formal records are to be typewritten on company stationery, minutes must be done for all status report or scope change meetings, and all documents must be controlled in accordance with the ISO 9001:2000 specification.

Communications Objectives

The five principal communications objectives to emphasize throughout the project are as follows:

- Project a positive image for KLSJ, the project, and the finished product.
- Ensure that both political support and financial support are forthcoming.
- Generate local interest in the Water Park to stimulate sales for the client during the first months of operation.
- Generate local interest in being an employee of the Water Park to secure labor for the opening of the facility.
- When necessary, anticipate and preempt possibly negative coverage from all potential sources.

These objectives are to be considered in all communications, especially in dealing with government, the media, or the general public. No opportunity for positive publicity should be overlooked or ignored.

Communications Planning

Communications will be managed as an internal project deliverable. All communications actions will be specifically identified in the Work Breakdown Structure (WBS) where applicable. The communications process will begin formally with the publication of this plan, which appoints the Team Leader as the communications coordinator (see "Responsibilities," below). The TL is to organize, implement, and control the published communications plan throughout the project. The TL must ensure that the implementation of the plan involves all relevant stakeholders, especially owners, investors, contractors, suppliers, regulators, and all team members.

The Communications Plan will be reviewed quarterly and at the start of each phase. The TL will collect feedback from all stakeholders. The feedback will be analyzed, interpreted, and used to improve the communications plan. The TL will document the lessons learned from this feedback for possible use in the later phases and in the Post-Project Report (see Chapter 16) for use on other projects.

Communications Messages

In dealing with government, the media, the general public, or any other external stakeholder, the following four major themes will be communicated:

> *Theme 1.* Ottawa is growing up. As a major world center and a national capital, it needs a water park (supporting statistics on Ottawa and water parks are to be made available to all section heads).
>
> *Theme 2.* This project is good for the local economy: It attracts tourists, which provides jobs for the local population, particularly for youths and students.
>
> *Theme 3.* A water park provides a healthy, wholesome family-oriented entertainment experience that can be enjoyed by people of all ages.
>
> *Theme 4.* This water park is environmentally friendly: All equipment is state of the art and energy-efficient, the water is purified and recycled, the facilities make use of existing structures, and the location is served by both bus routes and bicycle paths.

Because this project will involve major construction of several large structures, there is always the remote possibility of loss of life. Any initial media response to such a tragic event obviously should stress regrets at the incident but also point out the excellent safety record of KLSJ as well as that of the contracted construction firm. There should be no public discussion of responsibility or fault on the part of either the firm or the project until a thorough accident investigation has been completed. In the event of a major tragedy, it is expected that a media center will be established by the KLSJ head office to deal with media queries, press conferences, and any subsequent formal procedures.

Audiences

The Stakeholder Communications Chart shown in Appendix A depicts the various groups and individuals affected by this project. Because of the nature of the project, there are numerous stakeholders, both internal and external. The Project Manager (PM) and the TL must manage each stakeholder carefully to ensure that unexpected communications failures do not occur.

The principal internal stakeholders are the owners, the team members, and KLSJ headquarters. The principal external stakeholders are politicians, regulators, local investors, the local media, and local special-interest groups. Secondary external audiences are the general population and regional media that are not located in Ottawa but for which the project is potentially newsworthy. Appendix A describes how each type of stakeholder will be managed from a communications perspective.

Priorities and Constraints

The most important initial communications priority will be generating support from politicians, investors, and community groups. The client already has stated that the project will not proceed beyond the planning phase without secured investors and proper building permits. The National Capital Commission (NCC) must be given special attention because of its influence on construction in the Ottawa area.

Once the Water Park's site is selected and approved, ongoing communications with local communities (individuals and groups) must take place to maintain support and mitigate any potential protests or adverse publicity. See "Communications Objectives" and "Communications Messages" for more precise guidance.

Some private investors may wish to remain anonymous. This constraint will limit the potential benefit of name recognition in luring other investors. This can be offset by relying more heavily on corporate investment and sponsorship.

Responsibilities

The TL is responsible for implementing and monitoring the Communications Plan and will report directly to the Project Manager on all matters related to this plan. The TL will report personally on the communications plan at each end-of-phase meeting and at other times as described in the WBS.

The PM is responsible for external communications and is the only person authorized to speak with the media, politicians, or investors. The other section heads (finance, procurement, production, etc.) are authorized to communicate with their respective stakeholders and may make contractual arrangements within the limits set in the Procurement Plan.

Performance Reporting

The TL is responsible to the PM for all project performance reporting. Specific measurements, metrics, and standards are to be developed in accordance with the ISO 9001:2000 standard described in the Quality Management Plan. The PM, in consultation with the TL and the Quality Manager, will determine who is to gather, collate, and report on each performance objective. Performance reporting shall be carried out as described in the Stakeholder Communications Chart in Appendix A. Any change requests resulting from performance measurement will require that a form be completed and reported in accordance with procedures described separately.

Reporting of project results is through the TL to the PM. The PM is the only person authorized to release information related to milestones or phase completion. Results are to be distributed in accordance with the Stakeholder Communications Chart. When performance reporting identifies a need for preventive or corrective action, the PM or TL, as appropriate, will inform the responsible party in writing and monitor the results to ensure compliance.

Documentation

The TL is responsible for setting up and managing the document administration function in accordance with standard office procedures prescribed by KLSJ. The TL will ensure that project documentation, which is written purely to manage the project, is separated from client documentation, which is to be turned over to the customer upon project completion. A copy of all project correspondence, documents, drawings, and any other written communication originating from the project office is to be kept on file regardless of the originator.

All original permits, warranties, certificates, authorizations, financial obligations, leases, guarantees, and similar documents are to safeguarded off the job site, preferably at KLSJ headquarters. Copies of all such documents are to be retained at the project office for reference.

Administrative Closure

The TL, as office manager, shall comply with standard KLSJ administrative closure procedures. Emphasis will be on the following items (to be added to the WBS):

- Formal review and closure of all outstanding communication action items
- Formal post-project capture of lessons learned
- Formal project acceptance by the client, including signed documentation to that effect
- Formal handover of documents to the client
- Detailed archiving and storage of all project documents and supervised transportation from the work site to KLSJ headquarters

Other items, as deemed necessary by the TL or PM, may be added to ensure proper closure and thorough postproject archiving.

Appendix A: Stakeholder Communications Chart

The Stakeholder Communications Chart is shown in Table 6.2.

Table 6.2: Stakeholder Communications Chart

Who	How	What	When	Where	Why
Client (with the PM)	• Telephone • Informal and formal meetings • Formal documents (status reports and change requests)	• Progress • Problems, changes, or decisions • Acceptance	• Constantly, informally and formally • Monthly status reports • Project closure	• At the site • PM's office • Client's office	• To keep client apprised of changes and delays • To reassure client with plans to deal with the unexpected • To seek approvals
Investors (with the PM)	• Conversation (face to face) • Phone • E-mail • Letters	• Progress and status • Financial • Other investor issues	• Constantly and informally • Monthly status reports	• At the site • PM's office	• To keep them informed • To work together on financing and cash management issues
Politicians [with the PM and TM (Legal)]	• Meetings • Letters • Telephone	• Government support • Site lease • Progress and status • Problems and issues • Approvals	• As soon as site is selected • As issues arise • At least quarterly	• At the site • PM's office • Politicians' offices	• To arrange lease terms • To gain support and approvals as necessary • To advise on progress or problems
City inspector (with the PM)	• Telephone • Meetings if required • E-mail	• Building codes • Site inspection • Approval of building permit	• Prior to signing lease • As often as practicable prior to seeking approvals	• At the site • City hall	• To ensure there are no zoning surprises • To ease approval process

(Continued)

Table 6.2: Stakeholder Communications Chart (*Continued*)

Who	***How***	***What***	***When***	***Where***	***Why***
Fire marshal (with the PM)	• Meetings if required • E-mail • Phone	• Site inspection • Building codes	• Prior to signing lease • At the beginning and the end of construction • As required	• At the site • PM's office • Fire marshal's office	• To ensure building meets existing fire codes • Fire and safety inspection
By-law enforcement officer (with the PM)	• Telephone • Letters • E-mail • Fax	• Inform of upcoming project	• Before starting the project • When required	• Project manager's office • At the site	• To ensure proper access to the site for all personnel and contractors
Heritage groups (with the PM)	• Telephone • Meetings when required • Formal letters	• Inform of status of project	• When and if required	• Group's offices • Community meeting location	• To describe work and clarify issues concerning the project • To mitigate opposition
Chamber of Commerce (with the PM)	• Chamber of Commerce meetings • Phone • Letters • E-mail	• Look for business opportunities among members • Publicity	• Prior to park opening • On regular basis	• Chamber of Commerce's offices	• To promote the water park • To facilitate business opportunities

Suppliers [with the PM, retail expert, and TM (Finance)]	• Informal meetings • E-mail • Phone • Fax	• Discuss purchases • Delivery time • Changes in prices • Arrange payment	• Throughout project • Continuous contact	• PM's office • Suppliers' offices	• To ensure procurement of merchandise • Possible problems faced by the supplier
Advertising firm (with the PM and the marketing expert)	• Regular meetings • Telephone conversation • E-mail • Fax	• Promotion and advertising	• Prior to park opening • Regularly and informally	• Advertising firm's office • At the site • PM's office	• To establish an innovative marketing campaign for Ottawa–Carleton Water Park • To develop marketing plan
Contractors [with the PM and TM (Design)]	• Formal documents • Informal meetings as required • Telephone • E-mail • Fax	• Inform of upcoming project • Requests for proposals (RFP) • Walk-throughs • Weekly progress meetings	• Prior to release of RFP • As per work schedule • Prior to bid submissions • While performing work	• At the site • PM's office	• To describe work and clarify issues • To arrange contract • To update progress and voice concerns
Employment Canada (with the PM and human resources)	• Telephone • Letters • E-mail • Fax	• Hiring staff • Job openings	• Prior to advertising for help • When and if job openings arise	• Employment Canada's office	• For assistance in hiring staff • For possible part-time work during peak season

7

Risk Management Plan

Risk Management Plan: Discussion

A Risk Management Plan (RMP) attempts to identify risks that may occur during the life of a project, to measure the impact of those risks and make plans for handling the risks if they occur.

Risk management is a proactive and iterative process that is vital to controlling costs, meeting deadlines, and producing quality results. Depending on the complexity of the project, a risk management plan can be formal or informal. On small, straightforward projects, the identification of risks and the proposed actions could be done as part of the proposal or the project plan. For larger, complex projects, a separate RMP should be developed.

The RMP serves as both an internal and an external document. Although it is intended primarily for the members of the project team, it also serves as a key document for clients and stakeholders. Generally, it is fear of the unknown that raises concerns among clients and stakeholders. The RMP therefore is intended to demonstrate to the client and the other stakeholders that the project team has considered many of the things that could go wrong on the project and will manage those risks formally. The aim is to provide the stakeholders with a sense of security: The Project Team is prepared to handle any situation and minimize the impact to the project's schedule and budget.

The RMP usually is written by the Project Manager with advice and assistance from a risk management team. On large projects, that team will consist of the Project Manager, a client representative, and team members with expertise in risk management. On smaller projects, it may consist of the Project Manager and a client representative, possibly someone with expertise in risk management.

Risk Management Plan Outline

Executive Summary

This is an optional section that summarizes the major points in the document. It should provide an overview to readers who do not need to know all the details so that they can decide if they need to read the remainder of the document. At a minimum, it should identify the risks that have a high probability of affecting the success of the project. It also should identify the overall risk level of the project as a whole: high, medium, or low.

Background

If you feel that your readers are not familiar with risk management, you have an opportunity here to educate them about it. Begin with the purpose of risk management and the need for a risk management plan as outlined above. The terms *risk* and *risk management* may mean different things to different people; defining these terms may be useful.

Provide the readers with a brief history of the project risks related to the original high-level objectives and constraints. Some or all of this information may be contained in the Project Concept (see Chapter 1) or elsewhere, and so you may wish to be brief here. It is also reasonable to repeat the main points if they are still valid.

Define the fours steps of risk management: identification, analysis, response planning, and monitoring and control. State that they will be detailed in the following sections.

Risk Identification

In this section, describe specifically what process will be used to identify the risks associated with the project. Typically, the entire Project Team gets together to brainstorm potential risks that may occur on the project. This meeting also can include stakeholder representatives. All risks are identified, ranging from major items such as the withdrawal of a major investor and acts of God to more minor issues. The composition of the Risk Management Team should be outlined here, using an organization chart. The size and composition of the team depend on the size and scope of the project as well as the relative experience of the people involved.

Create a form or a log to record potential risks. This form also will allow ongoing tracking and updating of the risks as the project progresses. It is important to create a meaningful title for each risk that will describe each one uniquely. Each risk is assigned a sequential number. This is particularly useful for large projects that may last for many months and may have hundreds of associated risks. Some risk have a "trigger," an event and date when something happens or an action has to be taken. This has to be documented as well.

Risk Analysis

To assess a risk accurately, you must look at the probability of its occurring as well as the impact on the project if it does occur. Risks can affect any combination of project schedule, cost, and quality.

This section should describe the overall criteria that will be used throughout the project to rank probability and impact. A probability rank of low, medium, or high is assigned to each risk. The criteria should be established, but the percentages should be modified to satisfy the requirements of the individual project.

Similarly, the criteria to determine the impact on the project must be established. Again, these risks generally are categorized as low, medium, or high. Since the impact can be to time, cost, or quality, the impact category of low, medium, or high can be assigned separately to each of the three constraints or with a single ranking based on which criterion is considered most important.

To obtain an overall picture of the total risk of the project, a risk table can be created (see Tables 7.1 and 7.2 later in this chapter). These tables list all risks that have been identified in a matrix according to their individual rankings. Based on the location of most of the risks in this table, the Project Manager can determine if the overall project risk is low, medium, or high.

Risk Response Planning

A mitigation strategy should be developed to define one or more actions required to mitigate or minimize a risk. These action plans may eliminate the risk or reduce its probability or impact.

If you wish to follow formal risk response methods, you can categorize the responses as actions to be taken immediately, actions to be taken when the risk is imminent, or actions to be taken when the risk actually occurs—a contingency plan.

Record each response in the risk log.

Risk Monitoring and Control

Projects are seldom static. Risks change over the course of a project's life cycle. In addition, new risks arise during the course of the project and should be identified and evaluated. As a result, the Risk Log is a dynamic document that must be reviewed regularly.

Define the frequency for the Risk Management Team's meetings. That team must meet to review the risks and update the Risk Log. The frequency of the meetings depends on the individual project. Any changes in status or action should be discussed and documented for future reference.

Most important, if a risk event occurs, the response action should be exercised quickly. This action is noted in the log.

Finally, warn the reader that regardless of how carefully one plans for risks, completely unforeseen events may occur. Depending on the severity, the Risk Team may be called into an ad hoc meeting to react to the event. The RMP must allow for this eventuality.

Risk Management Plan: Example

KLSJ Consulting

January 1, 2003

Risk Management Plan

Ottawa–Carleton

Water Park

14 Palsen St., Ottawa, ON, Canada, K2G 2V8

Contents

Introduction

History

Recognizing and managing potential risks before they impact the project is an essential aspect of project management. Risk management is a proactive and iterative process that is vital to controlling costs, meeting deadlines, and producing quality results.

As part of the Carlington Aquatic Parks project, KLSJ Consulting determined that formal risk management must be done to predict and control the cost, schedule, and quality of the deliverables of the project accurately. The purpose of this Risk Management Plan is to detail what activities will be performed to manage the risks associated with the Ottawa–Carleton Water Park project.

Definitions

Risk: Negative project risks must be managed. Negative risk is any circumstance, which may or may not occur, that can adversely affect a project's cost or schedule or compromise its quality.

Risk Management: "the systematic process of identifying, analyzing and responding to project risk" *PMBOK Guide*, 2000 ed., p. 127.

Scope of Risk Management

Four Steps of Risk Management

Effective risk management involves the following four steps.

1. *Risk Identification*

   List all the possible risk items.

2. *Risk Analysis*

   Determine the probability of the risk occurring and then determine the impact of the risk:

   Can it delay the schedule?

   Can it raise the cost?

   Can it affect the quality of the product?

   Using these data, prioritize the risks and decide which risk items to address in the next steps.

3. *Risk Response Planning*

   Devise action items that can reduce the probability or impact of the risks. Establish a plan to mitigate the effects of the risks.

4. *Risk Monitoring and Control*

   Set up a procedure to monitor the previously identified risks and evaluate the probability and impact of each risk.

   Continuously monitor and identify any new risks that arise.

Conducting a risk assessment early in the project planning phase will allow KLSJ Consulting to evaluate the overall viability of the project in addition to identifying potential problems with the project schedule and/or budget. Early identification, quantification, and response development will allow the Work Breakdown Structure (WBS) to be updated to reflect the actual situation. Ongoing monitoring will help ensure that the Ottawa–Carleton Water Park project will remain on schedule and within budget.

Risk Identification

Potential risk events that may affect the schedule, cost, or quality of the Water Park project will be identified in brainstorming sessions with members of KLSJ Consulting and Carlington Aquatics [the Risk Management Working Group (RMWG)]. Each risk event will be assigned a sequential number and a meaningful title and will be described succinctly. This data will be recorded on a Water Park Risk Management Form (see Appendix A). These forms will create a Risk Log, a written log of ongoing risks and the steps that have been taken to mitigate them.

Risk Analysis

Assigning Probability and Impact

Again, using discussion among the appropriate parties, each risk will be assigned a ranking for the probability of its risk occurring and the impact if it does occur.

A probability rank of high, medium, or low will be assigned, using the Table 7.1 (the percentages are approximate).

Similarly, each risk will be assigned a schedule impact as well as a cost impact rank of high, medium, or low, according to the criteria shown in Table 7.2.

Table 7.1: Probability Criteria

Probability Rank	***Description***
High	More than 50% probability of occurring
Medium	Between 25 and 50% probability of occurring
Low	Less than 25% probability of occurring

Table 7.2: Impact Criteria

Schedule Impact	
Impact Rank	***Description***
High	Delay the opening of the park beyond July 1, 2005 (more than 6 weeks)*
Medium	Delay the opening of the park to between June 15, 2005, and July 1, 2005 (3 to 6 weeks)
Low	Delay the opening of the park to between May 24, 2005, and June 15, 2005 (less than 3 weeks)

Cost Impact	
Impact Rank	***Description***
High	Could add more than 20% to cost of project (more than Can$2,500,000)
Medium	Could add 10 to 20% to cost of project (between $1,200,000 and $2,500,000)
Low	Could add less than 10% to cost of project (less than $1,200,000)

*Opening after the July 1, 2005, date is economically unsound, and delays beyond this date probably will result in deferring the opening of the Water Park to the following year.

Risk Tabulation

Use of the risk assessment allows the creation of an overview of the riskiness of the project. By creating a matrix with the risk probability on the *Y* axis and the overall impact of the risk on the *X* axis and recording all the potential risks that have been identified, KLSJ Consulting will be able to identify the overall level of risk associated with the Ottawa–Carleton Water Park project (see Tables 7.3 and 7.4).

- Total project risk classification: high.
- Uncertainty with investors and funding and potential difficulty obtaining approval for the land and zoning make this a high-risk project.
- Total project risk classification: low.
- Once the issues of investors and zoning approval have been resolved, the overall risk associated with the project drops significantly. At the start of Phase 3, the project is assessed as low-risk.

Table 7.3: Risk Table at Start of Project (Prior to Phase 1)

Probability *Impact*	*LOW*	*MEDIUM*	*HIGH*
HIGH	7. Construction company has major financial difficulties. 8. Major investor withdraws from project.	3. Environmental assessment requires major mitigation.	1. Failure to secure investors. 2. Political groups delay zoning approval.
MEDIUM		5. Unable to find suitable operations manager.	
LOW	4. Delivery of attractions delayed. 6. Design not approved. 7. Delays in obtaining construction permit. 10. Construction delays due to inclement weather.	11. Post-trial modifications delay opening.	

Table 7.4: Risk Table at the Start of Phase 3

Probability *Impact*	*LOW*	*MEDIUM*	*HIGH*
HIGH	7. Construction company has major financial difficulties. 8. Major investor withdraws from project.		
MEDIUM		5. Unable to find suitable operations manager.	
LOW	4. Delivery of attractions delayed. 6. Design not approved. 7. Delays in obtaining construction permit. 10. Construction delays due to inclement weather.	11. Post-trial modifications delay opening.	

- The estimated cost of moving from the initial concept to the start of phase 3 is Can$513,550.
  - Initial expenditure of approximately $500,000 is a high-risk investment.
  - Remaining $11,500,000 is considered a low-risk investment.

Risk Response Planning

Using further brainstorming with the appropriate experts, one or more actions will be defined to mitigate the risk. These action plans may eliminate the risk, reduce its probability, or reduce its impact. One mitigation strategy is to define a contingency plan: If the risk event occurs, an action can be taken to reduce the impact. Another mitigation strategy may be to do nothing, to react only if the risk event occurs.

Risk Monitoring and Control

The RMWG and the appropriate office of primary interest (OPI) will meet biweekly to review the risks. The group will review each risk item by using the Risk Management Forms. Any changes in status or action regarding the risk item will be noted in the Date/Action field. Any impact as a result of the risk should be documented for future reference.

The Date/Status field should be updated to contain the following:

Open:	Use this status for all newly entered risk items.
Active:	Use this status if the risk event has occurred.
Mitigated:	Use this status when a risk item is being reacted to.
Closed:	Use this status when the risk does not need to be monitored any longer.

In addition, new risks that arise during the course of the project should be identified, evaluated based on these criteria, and recorded in the Risk Management Log. These newly identified risks will form part of the ongoing responsibility of the RMWG.

The RMWG will consist of the following core members:

- Carlington Aquatic Parks: Dan Milks, President
- KLSJ Consulting:
  - Karen Dhanraj, Project Manager
  - Scott Kennedy, Team Leader (Design and Construction)
  - Steve Jackson, Team Leader (Finance)
  - Laverne Fleck, Team Leader (Legal)
  - Jim Harris, Risk Manager

Other resources will be called on to assist on an as-needed basis. This working group will be responsible for coordinating and controlling all aspects of risk management for the Water Park project.

The RMWG will be responsible for all risk-monitoring activities. The group will meet monthly to review the risk forms and react appropriately. All changes will be published. Special meetings will be arranged to address emergency items as required.

Risk Database and Summary Report

Risks and their associated activities will be tracked by Jim Harris, the Risk Manager, on behalf of the RMWG. The Risk Manager will prepare a summary of the Water Park Risk Management Forms, including the potential impact and the current status (see Appendix B).

Lessons Learned

As the project progresses, some risks will disappear, others will appear, and most will go through several levels of reaction. This information will be added to a database of historical data that will be used to solve similar future problems.

Summary and Conclusion

Risk management is one of the keys to project success. Although the initial risk management process has identified many of the potential risks associated with the project, many new risks will continue to surface during the two and a half years of the project. The ongoing risk management process will continue to raise issues and potential risks throughout the project. The Risk Management Team must monitor the probability and impact of risks constantly and revise the project completion date, costs, and deliverable dates accordingly.

Appendix A: Water Park Risk Description Forms

The following exhibits are the key risk description forms.

Exhibit 7.1 Description of risk no. 1.

Water Park Risk Management Form

Risk No: 1	**OPI**: Team Leader/Finance
Title: Funding is not secured in time to meet project deadlines.	
Description: Approximately Can$12 million in debt and equity financing is required prior to the beginning of Phase 3 (Execution). With an uncertain investment climate, it is probable that it will not be possible to secure the necessary funding we need in time to progress to construction in the summer of 2004.	
Probability: High	
Level of Impact:	
Overall	High
Schedule	High
Cost	Low
Mitigation Strategy: Emphasize importance of investment during presentations to interested parties and Ottawa-area organizations. If investors are not forthcoming, approach National Capital Commission (NCC) and City of Nepean for assistance. If financing is not secured by the time site approval and zoning approval are awarded, halt the project and conduct a complete assessment of options. Do not proceed to Phase 3 without funding secured.	
Risk Monitoring: **Date/Status**:	

Exhibit 7.2 Description of risk no. 2.

Water Park Risk Management Form

Risk No: 2	**OPI**: Project Manger

Title: A political action group successfully petitions the NCC, resulting in delays or denial of site approval and subsequent delays in the start of Phase 3 (Execution).

Description:

There are a number of active organizations that oppose the construction of the Water Park in its current location. These groups, acting alone or together, could direct sufficient political attention to this issue to force the NCC into protracted community consultations. Community hearings and studies could delay the issuance of site approval by many months, making on-time completion of the project impossible.

Probability: High

Level of Impact:

Overall	High
Schedule	High
Cost	Low

Mitigation Strategy:

Hold preemptive meetings with all concerned parties to hear their points of view. Develop and implement a comprehensive communication plan to counter all valid arguments against the park.

Have a second, more aggressive plan to counter specific groups that actively lobby the NCC. If delay appears imminent, hire a government relations firm to assist with obtaining political approval.

If delay actually occurs, suspend all project activity and reassess options for completion by original deadline.

Risk Monitoring:

Date/Status:

Exhibit 7.3 Description of risk no. 3.

Water Park Risk Management Form

Risk No: 3	**OPI**: Project Manager

Title: Unfavorable environmental assessment requires major mitigation.

Description:

The result of the environmental assessment reveals information that may have a major impact on the viability of the project. Environmental concerns may result in significant conditions being placed on approval for the land and zoning approval from the NCC and the suburb of Nepean. These conditions may range from minor solutions to major and potentially costly remediation. The level of time, cost, and effort required to mitigate these environmental concerns may prove to be so demanding that the viability of the site will become questionable.

Probability: Medium

Level of Impact:

Overall	High
Schedule	High
Cost	High

Mitigation Strategy:

Conduct an initial review of the property to determine previous owners and users. Conduct an initial environmental assessment to determine potential obvious risks prior to committing much time, effort, and money to the process. If these initial studies do not reveal any significant risks, proceed with a detailed environmental assessment, completed by an independent contractor. Environmental damage resulting from the prior operation of the sewage treatment plant should be identified clearly and documented. The responsibility for the environmental cleanup of these prior contaminants should remain the responsibility of the NCC.

Risk Monitoring:

Date/Status:

Exhibit 7.4 Description of risk no. 4.

Water Park Risk Management Form

Risk No: 4	**OPI**: Team Leader (Design and Construction)

Title: Construction schedule is delayed owing to delays in the delivery of attractions.

Description:

Many of the attractions for the Water Park, such as the water slides, are highly specialized items. Most Water Park attractions are custom designed and fabricated for each individual park. There are very few firms that fabricate water park attractions, and since each attraction is custom designed and built, the producer has limited capacity. Any delays in production, not only in the production of our attractions but in that of attractions for other parks that may come before ours in the production cycle, may have a negative impact on the delivery of our attractions.

Probability: Low

Level of Impact:

Overall	Low
Schedule	Low
Cost	Low

Mitigation Strategy:

When selecting firms to fabricate the Water Park attractions, distribute the work evenly among a few producers. If one firm fails to deliver, at least the park will have some attractions delivered on schedule.

Ensure that there are minimal design difficulties with the attractions to avoid any unnecessary delays. Establish a milestone on the WBS to ensure that the attractions (e.g., slides) are ordered at the earliest possible date. The attractions are to be completed and installed prior to the end of the December construction. Contracts with the attraction manufacturers should contain a penalty clause stating that if attractions cannot be installed in the fall, as scheduled, the manufacturer will be responsible for additional costs to have the attraction completed in April or May, prior to the grand opening.

Risk Monitoring:

Date/Status:

Exhibit 7.5 Description of risk no. 5.

Water Park Risk Management Form	
Risk No: 5	**OPI**: Project Manager
Title: Inability to hire a suitable operations manager at a reasonable salary.	
Description: The ongoing success of the Water Park depends on the ability of the operations manager. Prior experience operating a similar facility is critical to the success of the park. There are very few water parks in Canada, and as a result, there are few qualified candidates. There are more qualified operators in the United States, but many of them are reluctant to move to Canada. Those who are prepared to relocate demand high salaries.	
Probability: Medium	
Level of Impact:	
Overall	Medium
Schedule	Medium
Cost	Low
Mitigation Strategy: Start recruiting for a qualified operations manager early in the process, at least one year prior to the opening date. Promote the opportunities and rewards of running a new water park and the desirability of being able to design the systems and controls. Promote the quality of life aspect of living in Canada and in the Ottawa region. Offer bonus incentives based on operating profit and possible stock options in lieu of an excessive salary. There will be greater motivation for the operations manager to succeed, and it will encourage the manager to remain in the position for a longer period.	
Risk Monitoring: **Date/Status**:	

Exhibit 7.6 Description of risk no. 6.

Water Park Risk Management Form

Risk No: 6	**OPI**: Team Leader (Design and Construction)

Title: Final consultant design is not acceptable to the owners and/or investors.

Description:

One or more major aspects of the detailed design are not acceptable to the owners and/or investors when submitted for final approval. There are significant changes to the overall plan to accommodate site conditions and technical obstacles. The design does not meet the owner's expectations or changes the operations and/or image of the park in a significant manner.

Probability: Low

Level of Impact:

Overall	Low
Schedule	Low
Cost	Low

Mitigation Strategy:

Design consultant will provide monthly updates to the management committee and inform it of any changes to the concept as a result of the design process. The Project Leader (Design and Construction) will maintain close contact with the designers to ensure that the owner's interests are represented, including participation in biweekly design meetings.

The design consulting company's contract should establish payment milestones based on completion and acceptance of design documents.

Risk Monitoring:

Date/Status:

Exhibit 7.7 Description of risk no. 7.

Water Park Risk Management Form

Risk No: 7	**OPI:** Team Leader (Design and Construction)

Title: Project may be delayed as a result of financial difficulties with the construction company.

Description:

The Water Park will enter into a contract with the construction company. If the construction company experiences financial difficulties, payments to subcontractors and suppliers will cease, resulting in construction liens being placed on the property and the facilities. Removing the original contractor and replacing it would prove to be extremely difficult.

Probability: Low

Level of Impact:

Overall	High
Schedule	High
Cost	Medium

Mitigation Strategy:

When selecting construction firms, ensure that the Project Management Team conducts a thorough review of their financial history by using services such as Dun & Bradstreet. The construction tender package should include the requirement that the successful firm be bonded. Both a performance bond and a labor and material bond from a certified bonding company should be required.

Risk Monitoring:

Date/Status:

Exhibit 7.8 Description of risk no. 8.

Water Park Risk Management Form

Risk No: 8	**OPI**: Team Leader (Finance)

Title: A major investor withdraws from the project after construction has started.

Description:

Once construction has begun and the attractions have been received, the project is financially committed for substantially the entire Can$12 million-plus project cost. The withdrawal of a major investor at this point would make project completion tenuous at best and also could jeopardize scheduled payments to our firm.

Probability: Low

Level of Impact:

Overall	High
Schedule	High
Cost	Low

Mitigation Strategy:

Insert penalty clauses and withholding fees high enough that investors are discouraged from withdrawing after the commencement of Phase 3. Once construction has begun, put a builder's lien on all structures to ensure preferential treatment in the event of bankruptcy. If a major investor still withdraws, cease all construction and freeze all project accounts pending completion of a full financial reassessment.

Risk Monitoring:

Date/Status:

Exhibit 7.9 Description of risk no. 9.

Water Park Risk Management Form

Risk No: 9	**OPI**: Team Leader (Design and Construction)

Title: Construction permits are not approved in time to complete construction before winter.

Description:

Construction cannot commence until all necessary permits have been obtained. These permits are subject to a number of approval levels within the city, any of which could request additional information or substantiation. Depending on the nature of the request, resubmission could delay the receipt of permits by several weeks, making it impossible to complete construction before winter weather makes construction impractical.

Probability: Low

Level of Impact:

Overall	Low
Schedule	Low
Cost	Low

Mitigation Strategy:

Brief officials at the city of the project and solicit their input into the design process at an early stage. The design consultant should develop a relationship with city officials during the life of the project.

Submit draft plans to applicable city offices for comment during Phase 2 and correct as necessary. If plans are still rejected, reassign all available design resources to this task until it is completed. If delays persist, halt all preparatory construction work and completely reassess all critical time lines to determine if the completion date is still viable.

Risk Monitoring:

Date/Status:

Exhibit 7.10 Description of risk no. 10.

Water Park Risk Management Form

Risk No: 10	**OPI**: Team Leader (Design and Construction)
Title: Inclement weather causes construction delays.	

Description:

Unseasonal weather or inclement weather causes delays in the construction process and prevents major construction activities (such as pouring concrete) from proceeding as scheduled. Typical construction schedules would allow for delays based on average days of work lost based on past seasonal experience, but extended, unforeseen periods of inclement weather would not be taken into account.

Probability: Low

Level of Impact:

Overall	Low
Schedule	Low
Cost	Low

Mitigation Strategy:

The construction company may be able to manipulate some of the scheduled activities to suit the weather forecast. Overtime should be considered when forecasts of inclement weather threaten crucial critical path activities.

If the construction schedule slips, construction should be scheduled on weekends and possibly holidays to make up for time lost to inclement weather.

Risk Monitoring:

Date/Status:

Exhibit 7.11 Description of risk no. 11.

Water Park Risk Management Form

Risk No: 11	**OPI**: Team Leader (Design and Construction)

Title: Post-trial modifications and repairs take more than the three weeks allocated in the schedule, causing the grand opening to be delayed beyond the May 21 long weekend.

Description:

There is a three-week window between the trial opening and evaluation and the grand opening celebrations. During this period modifications and repairs will be carried out on noted deficiencies.

Major unforeseen circumstances could require that significant modifications be made to the operating setup.

Major modifications could take longer than three weeks to identify, design, and implement.

Probability: Medium

Level of Impact:

Overall	Low
Schedule	Low
Cost	Low

Mitigation Strategy:

Involve the operators early in the process. The majority of testing and balancing of the mechanical equipment can be completed prior to the trial opening and evaluations. The operators should be present during the commissioning stage of the project and should sign off that the equipment is indeed in good working order.

Ensure that the operators have a detailed plan identified as a milestone on their payment schedules. This plan should be reviewed and approved by the owners prior to acceptance.

Risk Monitoring:

Date/Status:

Appendix B: Water Park Risk Management Summary

A summary of the Water Park risk management is given in Table 7.5.

Table 7.5: Water Park Risk Management Summary, Revised October 8, 2002

Risk	*OPI*	*Probability*	*Impact*	*Status*
1. Funding not secured	TL (Finance)*	High	High	Open
2. Political groups delay approval	PM	High	High	Open
3. Unfavorable environmental assessment	PM	Medium	High	Open
4. Delay in delivery of attractions	TL (Design and Construction)	Low	Low	Open
5. Unable to hire suitable operations manager	PM	Medium	Medium	Open
6. Final design not acceptable	TL (Design and Construction)	Low	Low	Open
7. Financial difficulties with construction company	TL (Design and Construction)	Low	High	Open
8. Major investor withdraws	TL (Finance)	Low	High	Open
9. Delays associated with construction permits	TL (Design and Construction)	Low	Low	Open
10. Inclement weather delays construction	TL (Design and Construction)	Low	Low	Open
11. Post-trial modifications exceed three-week window	TL (Design and Construction)	Medium	Low	Open

*TL: Team Leader; PM: Project Manager.

8

Quality Management Plan

Quality Management Plan: Discussion

A quality management plan is intended primarily for the members of the project team but is also a key document for clarifying and documenting quality issues and expectations with the client. Quality management is intended to provide the client with confidence that the project will satisfy the required quality standards (*conformance to specification*) and also meet the client's stated needs (*fitness for use*). Quality is one of the three possible constraints in the project's cost-quality-time trade-off equation. It is therefore very important to define properly the level of quality required as well as any limitations this imposes on the other two factors.

There are two principal aspects to quality: planning and management. Management in turn can be broken down into two components: quality assurance and quality control. Quality planning involves creating a sound plan that will enable you to meet the agreed-on quality standards for the project. Quality assurance is the system for ensuring that the plan is carried out; quality control involves activities such as monitoring, inspecting, and testing to ensure that products and services meet the required quality levels. See Chapter 13 for a more detailed discussion of the practical aspects of quality control.

Before writing the quality plan, you, as the Quality Manager (usually the Project Manager), should ensure that the quality needs required by the client are defined and agreed on. Based on these needs, the Project Team should create a succinct quality policy that expresses the quality objectives and standards of the project. This should be done with as much involvement of the client as possible. Depending on the size of the project, this document could be anywhere from one paragraph to a page in length.

In creating the quality policy, be sure to review the project *environment* for clues to the kinds of quality characteristics that must be built into not only the end product but the project management processes. The relevant environment could include the work site, the location where the product will be installed or used, the project plans, and any quality-related contracts and specifications. Look for anything that could become a quality issue, either positive or negative, that should be dealt with in the plan. At the same time gauge the risk and magnitude of these potential quality problems for cross-referencing to the Risk Management Plan (see Chapter 7).

Having considered and documented the quality management issues, the team should define the overall quality assurance and quality control methods to mitigate the threat of a quality failure. The plan should assign responsibility for each major aspect of quality to a specific person or position. The team also should determine any resource investments needed as part of the quality control framework, such as training, vendor certification programs, insurance against warrant claims, and special inspecting and testing equipment. These investments must be built into the project cost estimate.

One last point regarding the Quality Management Plan: It is important to have a quality system that is documented, verifiable, and repeatable. The team and the client should be confident that if the system is followed, the proper level of quality will result. There should never be surprises with respect to quality. One method of ensuring a predictable degree of quality is to adopt a quality standard such as the ISO 9001:2000 series or a Total Quality Management (TQM) type of model using project management methods documented in the *Guide to the Project Management Body of Knowledge (PMBOK Guide)*. Many organizations have found these standards very useful in defining how to achieve quality results and educating their employees and suppliers about the procedures to ensure success. There are many superb books and manuals on the subject of ISO 9001:2000 and TQM, which the reader should seek out for more detailed information.

Quality Management Plan Outline

Executive Summary

This is an optional section that summarizes the major points of the document. It should provide enough detail that the reader can decide whether to read the remainder. At a minimum, it should highlight key quality issues and how they will be dealt with as well as the quality philosophy adopted for the project.

Background

Provide the readers with a little history because they may not know anything about the project. State how the project started, why, the original high-level objectives, the ballpark cost, and schedule milestones. Some or all of this information may be contained in the Project Concept (Chapter 1), the Project Plan (Chapter 5), or elsewhere, so you may wish to be brief.

Standards (ISO or Other)

In this section describe specifically which quality standard or standards will be used, for example, ISO 9001:2000 (not just ISO 9000, which is an overview document). Be sure to state whether you are registered or certified for this particular standard (if applicable) or if certification is pending. Other relevant standards also should be mentioned, especially in areas where there is the possibility of confusion. These areas could include jurisdiction of building codes, fire and safety codes, municipal by-laws, and similar quality-related regulatory standards. It is the Project Team's responsibility to ensure that it uses quality standards applicable to the work site or end user.

Organizational Structure: The Project Team

Describe the overall quality organization within the project. List the names of the individuals who are responsible for each aspect of quality as well as any specific delegated authority they have been given. If the actual names are not known yet, state the knowledge level required for each position. Such detail is important both internally and externally as a point of reference for deciding who has authority over which quality issues.

Diagram the quality organization within the Project Team, as shown in Figure 8.1.

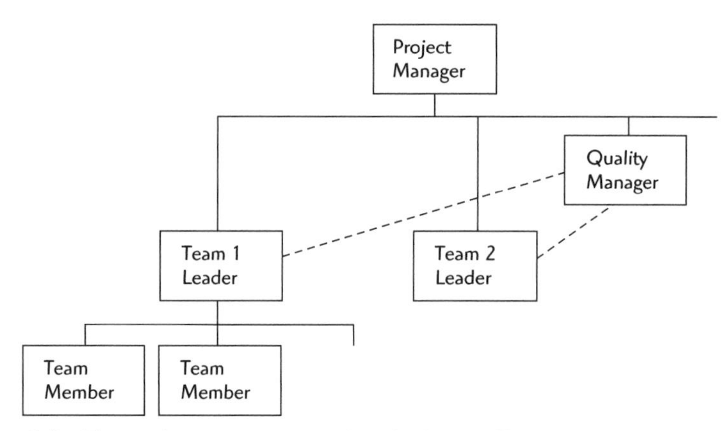

Figure 8.1 The quality organization within the Project Team.

Responsibilities

Detail the quality responsibilities of each member of the project. Here is an example:

Project Manager: Manages the quality team (leader, motivator, etc.).

Responsible for all external communication (quality planning, quality reporting, quality meetings, user and upper-level management interface)

Major Goal: Successful project (plan, control, communicate)

Quality Manager: Depending on the scope of the project, the Project Manager may appoint a Quality Manager (QM) and delegate the following:

Point of contact for all quality-related issues

Responsible for overall quality plan implementation, quality assurance, and control

Major Goal: Compliance with quality plan and standards

Team Leader: Supervises implementation of quality plan within the team.

Responsible for (not necessarily does) most quality management activities

Major Goal: Technical quality of the product

Team members: Responsible for quality in the construction of a specific item and for attending meetings, reporting errors or omissions, providing suggestions, and so forth.

Major Goal: Task completion within specifications

Mention any liaison with headquarters and other functional organizations. Detail who reports to whom, with dotted- or solid-line authority. You also may want to refer to other documents that specifically mention quality responsibilities.

Procedures

Describe the procedures that will be used to manage quality, such as product design, inspection, testing, contract review, document control, and supervision of workmanship in general. If these procedures are well documented elsewhere, it may be sufficient to reference those other publications. Otherwise, provide enough detail that team members with quality responsibilities know what needs to be done. Since this tends to be rather lengthy, an Appendix to the Quality Management Plan may be best; this could take the form of a manual of project procedures for each critical quality area.

Processes

Most activities within the project can have an impact on quality and must be managed to some extent. However, certain processes should be given special attention. Depending on the nature of the work, accuracy or structural integrity could be the key, or it could be ease of use or conformance to standards that is most important. Whatever the project, the Quality Manager needs to define the desired results and then identify which quality management processes will ensure success. This is a fundamental step in quality management because in some cases, particularly in smaller projects, it is not cost-effective or necessary to manage every aspect of quality on a daily basis. Identifying and closely monitoring a few critical processes can achieve the majority of the project's quality goals.

Resources Required

In some instances it may be difficult to separate the cost of quality management from the cost of project management in general. However, every effort should be made to identify the specific costs of additional personnel, equipment, facilities, and services that are necessary to achieve the level of quality desired by the client. For instance, special one-of-a-kind needs may require the purchase of unique testing equipment or the subcontracting of unique services. Prototypes may be needed or trial runs may have to be done to ensure conformance to standards and proper operation. Raw materials from suppliers may need to be inspected before use, and a certain amount of finished product may have to be subjected to destructive testing. The results of these tests and all quality activities have to be documented properly and stored for future reference.

The total of all these activities can be substantial; this is why it is important for them to be itemized. To make more informed decisions about the potential trade-offs between cost, quality, and time, the client should be aware of the resources needed for quality management.

Quality Assurance

Quality assurance is intended to provide the client with confidence that if the plan is implemented properly, the project will satisfy the specified quality standards and meet the client's stated needs. In the Quality Management Plan it is usually sufficient to identify which policies and processes will provide quality assurance to the client for each main quality specification or feature. More is said about the practical aspects of quality assurance in Chapter 13.

Quality Control

Quality control involves monitoring the actual processes, as opposed to quality assurance, which involves monitoring the implementation of the plan. For the purposes of the Quality Management Plan, it is generally sufficient to identify the processes and products that will be monitored and the measurements or results they will provide. As an example, to monitor project progress, Earned Value reporting may be done. Also, inspection of supplied materials may be done, and an unacceptably high reject rate could lead to decertification of that supplier (one possible outcome). More is said about the practical aspects of quality control in Chapter 13.

At the end of this section it may be useful to tie together the main activities in quality planning, quality assurance, and quality control into a table for the client to review. The table should show the client's principal needs, followed by a brief description of how each aspect of the quality management plan meets those needs. This process also may be used to find any gaps in the quality plan or any activities that are unnecessary.

Reports

There will be a need for periodic status reports as well as summaries of the results of key quality milestones as they are achieved. The format, content, and frequency of status reports, milestone reports, and other project quality documents should be stated in general terms. Also, this section should specify who receives each report and what his or her responsibility is after receiving it. A schedule may be useful for larger, more complex projects. See Chapter 13 for more details on quality monitoring and reporting.

Documentation

As with the overall project, there are user documents and project management documents. Describe which quality-related documents will be produced (for example, this plan), which ones will be turned over to the client, when they will be produced, and who is responsible for producing and storing them. Once again, a schedule or table may be a useful memory jogger for reference during and after the project.

Management Approval of the Plan

The Quality Management Plan should be reviewed and approved by the client, although the level of formality may be less than that for the overall project plan. If required, leave space for approvals.

Quality Management Plan: Example

KLSJ Consulting

December 1, 2003

Quality Management Plan

Ottawa–Carleton

Water Park

14 Palsen St., Ottawa, ON, Canada, K2G 2V8

Executive Summary

The attached Quality Management Plan describes in detail the steps that will be taken to ensure compliance with all quality-related specifications, requirements, and laws. Since the client has stated that safety and reliability are high priorities, the Project Team has emphasized these aspects in the overall plan. In addition, because of the size of the project, internal project quality also will be given considerable attention.

KLSJ Consulting will be basing its quality plan on strict adherence to ISO 9001:2000 standards for both quality control and quality assurance. Since the ISO standard was part of the KLSJ bid, no additional resources will be required to achieve this level of process and product quality. KLSJ has estimated that roughly 250 person-days will be devoted to quality management, control, and assurance. A sample ISO 9001:2000 Quality Procedure is provided in Appendix A.

A number of areas are highlighted in the plan as requiring special attention from the Project Team. Primarily, these areas are related to design, contracting, purchasing, testing, and document control, although all ISO procedures are covered thoroughly. In particular, KLSJ will keep detailed quality-related records to assist the client in securing the necessary city approvals and affordable liability insurance for the ongoing Water Park operation.

Status reports are expected to be frequent, in keeping with the client's wishes, but quality issues will be reported only on an as-required basis as well as at each significant milestone. The client will be consulted on all significant quality-related issues and changes.

This document constitutes a formal plan. It will be briefed to the client and must be approved and signed off by the client or the client's representative. Once accepted, the Quality Management Plan cannot be amended except through a formal change control process.

Contents

Background

Research has determined that a market exists in the Ottawa area for a full-scale water park complete with water slides, a wave pool, picnic facilities, and related summer attractions. KLSJ Consulting has been asked to manage the design, construction, and initial testing of the facility, which is to be conducted as a turnkey project. The estimated cost of the facility is Can$12,450,000, which is a Class B estimate (can be from +25% to −10% in error).

Owing to the fact that injury or even loss of life may result from a malfunction of the equipment, the owners have made quality a high priority for this project. Because of the large scale of the construction site and the amount of concurrent activity that must take place, quality must be well planned and closely monitored. There is a need to ensure safety at the work site during construction and provide a high degree of confidence in the workmanship. The ability of the owners to obtain liability insurance at a reasonable price will depend to a great degree on the quality of the facility that is constructed.

Standards

KLSJ Consulting is an ISO 9001:2000–certified company and uses the ISO 9001:2000 standard for all project management contracts. KLSJ is also fully compliant with the ISO 14001 standard for environmental management and responsibility, although certification in that area is pending. Other relevant standards that KLSJ will use to manage quality include the Ottawa-area building codes, fire safety standards, and workplace health and safety standards, as well as applicable municipal by-laws.

Organizational Structure

As shown in Figure 8.2 the organizational structure of KLSJ Consulting includes a quality system section within its headquarters. These people are available to the Project Manager (PM) to conduct training, audits, quality assurance, and similar functions. Although not part of the formal project team, they are considered on-call resources that the PM can task to assist with quality-related issues.

In addition, ISO 9001:2000, section 4.1.2, mandates the PM to appoint a member of the project team who, irrespective of other duties, shall be the project authority and point of contact for all aspects of quality. For the Ottawa–Carleton Water Park, that responsibility has been delegated to the Team Leader (TL) for Finance.

Responsibilities

All personnel at KLSJ have a responsibility for quality in accordance with ISO 9001:2000, section 4.1.2. Overall project responsibility for quality

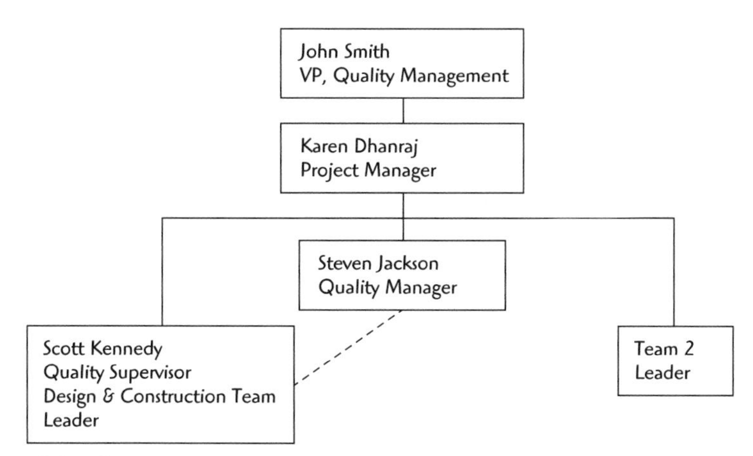

Figure 8.2 Quality System Section organizational structure.

rests with the PM notwithstanding the defined authority and responsibility delegated to other members of the Project Team. At KLSJ itself, the Vice President (VP) for Quality Management exercises corporate-level decision making with respect to quality and has the authority to negotiate with the client if there is a quality-related dispute.

Procedures

All quality control and quality assurance procedures are in accordance with the ISO 9001:2000 standard. Additional amplification (where necessary) will be contained in the Manual of Project Procedures (MPP), which the PM will complete prior to project start and keep updated throughout the project (see sample ISO procedures in Appendix A).

Processes

Detailed quality processes are based on the criteria of ISO 9001:2000, section 4.9, "Process (Work) Control," and will be contained in the MPP when completed. In addition to section 4.9, the following sections will be given particular attention throughout the project:

Section 4.3: Contract Review—for the client contract and all subcontractors

Section 4.4: Design Control—for all architectural and structural design

Section 4.6: Purchasing—to ensure that all supplied material is in accordance with specifications

Section 4.10: Inspection and Testing—to ensure that all specifications have been satisfied

Section 4.16: Control of Quality Records—to demonstrate to the client and any third parties that all aspects of the project were completed in compliance with quality standards and specifications

Resources Required

Quality is an integral part of operations at KLSJ Consulting, and so most of the quality resources already have been factored into the bid price and cannot be separated readily. Based on rough order-of-magnitude estimates for projects of similar size, we predict that approximately 250 person-days will be required to perform quality planning, control, and assurance. Additionally, resources will be required to manage the related inspection, verification, record keeping, and training of all initial staff on service quality as they relate to a water park environment.

Quality Assurance

Quality assurance ensures that the quality processes are implemented. Quality assurance for the Ottawa–Carleton Water Park project will be managed primarily through the ISO 9001:2000–mandated activities, with particular attention to the following:

- Design control (incorporation of required standards and specifications)
- Contract review (compliance with stated standards and specifications)
- Document control (quality records with respect to performance and compliance)
- Verification of purchased product (to ensure conformance to standards and specifications)
- Control of nonconforming product (documentation and quarantine of any product deemed not to be in accordance with standards or specifications)
- Corrective and preventive action
- Internal quality audits (performed by an independent representative from KLSJ headquarters at no cost to the client)

Specific quality assurance verification will be based on requirements described in the quality planning process. Methods of providing assurance are shown in Table 8.1. The client should note the following additional KLSJ policies with respect to quality assurance and workplace safety:

- All material must be at least contractor grade.
- All equipment must be certified and approved by the safety standard(s) applicable to the jurisdiction of the project.
- The premises will be inspected prior to the commencement of construction and at regular intervals for any unacceptable workplace hazards, including deficient fire prevention and protection systems, inadequate ventilation, and structural inadequacies.
- All construction personnel must abide by strict worker safety standards. This includes zero tolerance for any practice that endangers workers' safety or the safety of others (this includes cutting corners, using substandard equipment or materials, and being under the influence of any substance while on the construction site).

Quality Control

The construction and engineering drawings will identify in detail the specifications and standards for the project (see Table 8.1, for examples). Quality control of these requirements will be carried out in accordance with the ISO 9001:2000 standard, with particular attention to the following:

- Section 4.4.7: Design Verification—to ensure that the Water Park design complies with specifications, codes, and so on
- Section 4.4.8: Design Validation—to ensure that the design is in keeping with the wishes of the client and investors

- ▲ Section 4.6.2: Evaluation of Suppliers—to monitor and control the quality of any product or service purchased from or subcontracted to a third party that can have an impact on the overall quality of the project
- ▲ Section 4.6.4: Verification of Purchased Product—to confirm prior to use that delivered goods were in fact those selected and purchased by the PM or the client
- ▲ Section 4.10.2: Receiving, Inspection, and Testing—to confirm prior to use that materials conform to standards (this will be done using recognized industry methods)
- ▲ Section 4.13.2: Review and Disposition of Nonconforming Product—to identify and quarantine nonconforming product until it can be disposed of, reworked, or otherwise certified for use by a duly authorized representative of the Project Team (normally the TL)
- ▲ Section 4.16.1: Control of Quality Records—to ensure that documented proof of compliance with standards and specifications is captured and stored for future reference, and handed over to the client upon completion of the project

As an illustration of how KLSJ Consulting manages for quality, Table 8.1 shows three potential quality objectives and the related activities that would be undertaken to ensure total customer satisfaction.

Reports

Formal quality status reports will be provided at dates and times specified in the Work Breakdown Structure (WBS), and as agreed on with the client and other stakeholders. Normally, KLSJ will report quality issues by exception (that is, when there is a quality problem), but quality reporting also will be part of all milestone and end-of-phase meetings.

The quality reports always will be sent to the client but also may be sent to other stakeholders for information or action when there is a need to know. It is KLSJ's responsibility to document each quality issue and make recommendations on a course of action. It is the client's responsibility to approve the course of action.

Documentation

The project staff and contractors will require *management documents* such as the Quality Management Plan and the results of quality studies and quality control reports during the project. Additional user documentation such as quality maintenance of the site will be produced as described in the WBS. All project documents will be made available to the client and the relevant operations staff at specified milestones or at the end of the project. Contractors will be provided with copies of studies and relevant information as required.

Table 8.1: Quality Objectives and Related Activities

Quality Objective	*Quality Planning*	*Quality Control*	*Quality Assurance*
All construction will be carried out in accordance with the Ottawa-area building code	Include the following activities as part of the Work Breakdown Structure (WBS): • Building inspection • Construction design • Building permit • Postconstruction inspection	Take the following actions as required: • Monitor materials used • Observe methods • Clearly specify requirements in design	Document the following activities for the client: • Building permit approval • Client approval of final construction
All rides are in good working order and safe for use	• Client defines safety standard and level of effort • Manufacturer specifies how rides are to be built and used	• Follow manufacturer's instructions and specifications • Independently check all critical assemblies	• Insert detailed verification requirements into the acceptance testing regime
Staff are trained and knowledgeable about customer service and amusement park operations	• Client specifies staff quality needs as part of quality policy • Training is included in the WBS • Hiring and training are done according to a specific plan and schedule	Human resources department of KLSJ will do the following: • Approve the hiring and training plan • Monitor all training given • Appoint a member as part of the hiring committee	• Test staff knowledge at the end of training • Spot-check skills and knowledge during the trial run

Management Approval of the Plan

This plan will be briefed to the client and KLSJ management at a point to be specified in the WBS. Formal approval and sign-off will be required. Once accepted, this Quality Management Plan becomes a key source document for the project and will be safeguarded as such. Subsequent amendments will be permitted only through a formal quality review process.

Appendix A: Manual of Project Procedures

KLSJ Consulting ISO 9001:2000 Sample

Section 4.11: Control of Inspection, Measuring, and Testing Equipment

KLSJ Consulting ISO 9001:2000 Manual of Project Procedures	
Approved by:	(VP Quality)
Implemented by:	(Project QM)

1.0 Purpose

1.1 The purpose of this section is to define a policy that will ensure that all inspection, measuring, and testing equipment utilized on the Water Park project is adequately controlled, calibrated, and maintained at a level that conforms to industry and KLSJ quality standards.

2.0 Scope

2.1 This quality policy applies to all inspection, measuring, and testing equipment in use at the Water Park project site whose use directly affects quality.

3.0 Flow of Work

3.1 Control will be achieved through a three-step work flow process involving recall, calibration, and recording of information. Table 8.2 provides details.

Table 8.2: Work Flow Process

Serial	*Action*	*By Whom*	*Procedure*
1	Recall	Calibration section, KLSJ headquarters	On or before its due date, the calibration section will recall all inspection, measuring, and testing equipment that is due for calibration that month. A Test Equipment Transit and Receipt form will be issued to each project that is required to submit equipment for calibration. The Project Manager (or a delegated representative) will forward the equipment to the calibration section. Note: Equipment will not be signed out to a project if it is due for calibration in that month or the next month until it has been calibrated according to this policy manual.
2	Calibrate	Calibration section	The Test Equipment Maintenance Management Information System (TEMMIS) lists the authorized procedures for each piece of equipment. This procedure number will be annotated on the work order and referred to during calibration. All calibration shall be performed in accordance with the approved technical manual or published procedure. Upon completion, each calibrated piece of equipment will be identified with a calibration seal noting the date and identity of the technician. Seals will be placed in a manner to show evidence of tampering if the equipment has been opened after calibration.
3	Record	Calibration section	The data on the completed work order will be entered into the TEMMIS database. The calibrated equipment will be returned to the applicable project (or to central stores if the equipment has been replaced).

4.0 Responsibilities

4.1 The KLSJ Calibration section is responsible for recalling and calibrating all KLSJ-owned inspection, measuring, and testing equipment that may affect process quality.

4.2 Calibration records and scheduling for each individual piece of equipment are maintained by the Test Equipment Maintenance Management Information System database. Each piece of equipment is assigned a permanent identification number for tracking purposes.

4.3 Once a recall has been issued, the Project Manager will ensure that each piece of equipment utilized in the project is returned for calibration on or before the applicable due date. The Project Manager also may return equipment for calibration whenever equipment accuracy is suspected of being compromised.

5.0 Procedures

5.1 All calibration carried out by the calibration section shall be performed in accordance with the approved technical manual, manufacturer's manual, or local procedure. The procedure authorized for each piece of equipment is identified in the TEMMIS database.

5.2 Each piece of equipment that has been calibrated by the calibration section will be identified with a calibration seal. This seal identifies the date of calibration, the due date for the next calibration, and the technician who carried out the most recent calibration. Seals will be placed in a manner to show evidence of tampering if the equipment has been opened subsequent to calibration.

5.3 When the equipment cannot be calibrated by the calibration section because of pending repair or maintenance, the equipment will be appropriately tagged and quarantined in accordance with section 4.13, "Control of Nonconforming Product."

6.0 Document History

6.1 This procedure was issued on March 21, 2003. It was revised on February 13, 2004. The OPI is Director of Logistics, KLSJ Headquarters.

Process Owner Sign-off
Reviewed and Accepted by (name and signature):

9

Procurement Plan

Procurement Plan: Discussion

A procurement plan describes how consultants, contractors, and suppliers will be selected to provide goods or services for a project. Working with the client, the project manager must determine the overall procurement strategy for the project. This choice is governed by budgets, time lines, political considerations, and policies of the client or the project management firm.

A procurement plan serves as both an internal and an external document. Although it is intended primarily for the members of the project team, it may be provided to clients and key stakeholders. The plan is intended to provide the client with confidence that the goods or services will be obtained for the best value for money based on the triple constraint of time, cost, and quality.

Depending on the complexity of the project, a procurement plan can be formal or informal. On small, straightforward projects, the identification of the procurement strategy can be done as part of the

Proposal or the Project Plan. On larger, complex projects, a separate procurement plan should be developed.

Before drafting a procurement plan, the author should have a thorough understanding of the project's environment, the process of procurement, and any procurement policies of the client that may limit or influence that process. For example, one client's procurement policy may state that all contracts over a designated value must be bid competitively, whereas another client may give preference to local suppliers.

In creating the procurement plan, be sure to review the project environment to assess any limitations or constraints that may affect the procurement strategy and evaluate the impact on the project's schedule, quality, and budget. Generally, a competitive bidding process will ensure the best price for the work, although this requires significant preparation and can be time-consuming to implement. Since directed contracts limit competition, they may lead to higher costs, although there may be some advantages with respect to time and quality.

Depending on the size of the project, the Procurement Plan can be anywhere from one page to an entire volume in length. A project may use multiple types of procurement policies and contracting arrangements. Each contract should be evaluated separately to determine the most effective contracting method. At the same time, one should gauge the risk and magnitude of each contract and determine the overall risk to the project, as detailed in the Risk Management Plan (see Chapter 7).

Procurement Plan Outline

Executive Summary

This is an optional section that summarizes the major points of the document. It should provide an overview to readers who do not need to know all the details that they can decide if they need to read the remainder. At a minimum, it should identify the overall procurement strategy that will be used on the project and highlight key contracts.

Background

Provide the readers with a brief history of the project and the original high-level objectives and constraints. Some or all of this information may be contained in the Project Concept (Chapter 1) or elsewhere, and so you may wish to be brief here. It is perfectly acceptable to repeat the main points if they are still valid.

Objectives

In this section, describe the overall objectives of the Procurement Plan and the specific procurement strategy that will be used. For example, it can be stated that all aspects of the project will be bid competitively. Other relevant information, such as a list of all the major contracts that require procurement, also should be mentioned.

Organizational Structure: The Procurement Team

Describe the overall organization of the procurement team for the project. List the names of the individuals who are responsible for each aspect of procurement. If the names of individuals are not known, identify the positions or skill sets that are required.

Outline the responsibilities of each member of the procurement team and what each one is responsible for at each step in the process. Detail any reporting relationships, such as who reports to whom, and identify any specific delegated or decision-making authority. You may want to identify clearly specific people or individuals who will be responsible for liaison with the client and/or the service providers.

Procurement Strategy

The procurement strategy describes the procedures that will govern all aspects of the procurement process. General conditions, such as "all contracts will be competitively bid," are elaborated on and clarified to suit the size and scope of the project (e.g., contracts over Can$5,000 must have a minimum of five qualified bidders). Any conflict between the client's and the plan's procurement policies should be clarified here.

This section also identifies who will be signatory to the contract (the client or the project management firm) and who has authority to grant exceptions to the procurement policies. It earmarks the individuals who will have the ongoing authority to administer the contract on a daily basis and details the responsibilities and authorities granted to those individuals.

Scope of Activities

Identify all the tasks that will be procured from independent suppliers, consultants, and contractors. This should be an exhaustive list of all procurement activities on the project. Along with each of these tasks, describe the type of contractual arrangement that is being proposed and the estimated costs for service. Provide enough detail so that the project team members will know what incentives and/or limitations are associated with each contract; for example, indicate bonus and penalty clauses for early and late completion.

Procurement Schedule

Detail the procurement activities, their estimated time lines, and their duration. Often this information is contained in the project's Work Breakdown Structure (WBS). This section does not have to repeat all the contents of the WBS, although there should be some reference to where the information is available for future reference.

Risk Assessment

To assess the risk associated with each of the procurement contracts accurately, you must look at the probability of a risk occurring as well as the impact to the project if it does occur. Risks can impact the project schedule, the costs, or both. This section should describe the overall criteria that will be used throughout the project to rank probability and impact. A probability rank of low, medium, or high is assigned to each risk. The criteria should be established, but the percentages should be modified to satisfy the requirements of the individual project. Similarly, the criteria for determining the impact on the schedule and cost must be established. Again, these risks generally are ranked as low, medium, or high. Complete details of the risk assessment are outlined in the Risk Management Plan (see Chapter 7).

Project Review

To ensure clear communications and uniformity of standards, this section details the procedures and tools that all service providers will use to communicate with the project management team and with each other. Detail any specific standards, software, and scheduling tools that may be required and apply these requirements uniformly and consistently across all procurement contracts. Similarly, ensure that all contracts include a commitment to attend and participate in regularly scheduled project status meetings.

Related Project Plans and Documentation

Describe any project documentation or management reports that must be submitted, including the frequency and content of each submission.

Terms and Conditions

Standardized terms and conditions are prepared by industry associations in conjunction with owners and service providers. These standard documents provide security for both parties in that they are based on industry standards. Owners are protected by clauses governing the quality of the work or service provided, and service providers have some protection from having payment withheld unreasonably.

Assumptions

Do not force the readers to make their own assumptions. Clearly state the procurement team's assumptions. Information may become available at a later time that can affect or change the procurement requirements, and so it is beneficial for readers to know what the intention was at the time of writing. Statements should be unambiguous, with the emphasis on facts.

Approvals

Regardless how carefully the project is specified or the procurement process is planned, there will be unforeseen circumstances that will require that activities deviate from the plan. To maintain control and accountability, any deviation should require written approval.

Clearly identify the signing authorities or levels of approval required to implement a change. This process should parallel that described for the change control section in Chapter 14. Identify the people (or positions) who have authority, describe the authorization or responsibilities they may approve, and indicate constraints or limiting factors associated with the signing authority.

Procurement Plan: Example

KLSJ Consulting

January 1, 2003

Procurement Plan

Ottawa–Carleton

Water Park

14 Palsen St., Ottawa, ON, Canada, K2G 2V8

Contents

Appendixes

Appendix A: Architectural Design Detailed Procurement Schedule

Appendix B: WBS and Procurement Schedule for Detailed Architectural Design Services

Appendix C: CCDC and CCA Standard Construction Contracts

Background

Dan Milks, president and CEO of Carlington Aquatic Parks of Ottawa, has identified the *market need and potential benefits* of a water park in the National Capital Region. Mr. Milks envisions a family-oriented water and amusement facility that would provide a recreational outlet for visitors and local residents in a city lacking this type of outdoor recreational attraction. Mr. Milks issued a *Project Concept* document in September 2002 and began the process of marketing his idea to the local government and investors. The concept is based on a spring 2005 opening for a water recreation facility in the Ottawa area costing about Can$12 million with a further $4 million available for future development.

The size and cost of the project guarantee that there will be a large number of organizations and individuals directly involved or indirectly interested in the development of the park. A list of key *stakeholders* includes the following:

- Owners
- Investors
- Government: provincial, regional, and municipal, including the City of Nepean, the National Capital Commission (NCC), the Regional Municipality of Ottawa Carleton (RMOC), and the Ontario Municipal Board
- Community groups
- Contractors
- Employees
- Related business partners
- Competitors

Given the substantial support throughout the region for the idea,[1] Carlington Aquatic Parks has decided to proceed with the development of the Water Park project. KLSJ Consulting has been contracted to manage the project. Since a large portion of the design and construction work must be procured from outside sources, KLSJ has developed the following *Procurement Plan* to map out the process for the successful and timely completion of the Ottawa–Carleton Water Park.

Objectives

The overall *objective* of this Procurement Plan is to describe how KLSJ Consulting will select consulting and construction firms to complete the required studies, the design, and the construction of the Ottawa–Carleton Water Park facility on behalf of Carlington Aquatic Parks. The Water Park will be completed and ready for a scheduled opening on May 21, 2005, at a total cost of Can$12,450,000. Note that this cost is a Class B estimate with a range of +25% to −10%.

Project management theory[2] speaks of a *triple constraint*: time, cost, and quality. Given the fixed target date for the opening of the park and the low probability that extra funding could be located on short notice, any project-related problems probably will result in a decrease in the number of attractions or their quality. KLSJ Consulting proposes that all aspects of the project be bid competitively to ensure that Carlington Aquatic Parks receives the work at the best possible price.

The major contracts that will require procurement include the following:

- Initial architectural design
- Market research
- Environmental study
- Archaeological study
- Traffic study
- Site services study
- Environmental assessment*
- Detailed architectural design
- Design engineers
- Construction management firm
- Initial architectural design

Procurement Team

The *Procurement Team* will consist of three core members for the duration of the project, plus additional personnel with expertise in specialized areas, to be called on as required for evaluation of selected areas of the design proposals. The basic organization of the team is shown in Figure 9.1. Close horizontal communications among the team members will be essential.

---

*Since the initial approval of the Procurement Plan, Carlington Aquatics has approved the sole source selection of the Environmental Assessment contract. Owing to slippage in the project schedule, it was decided that the firm that conducted the initial Environmental Study (which was selected originally through a competitive process) would be awarded the Environmental Assessment contract. Awarding the contract on a sole-source basis will allow the project to get back on schedule.

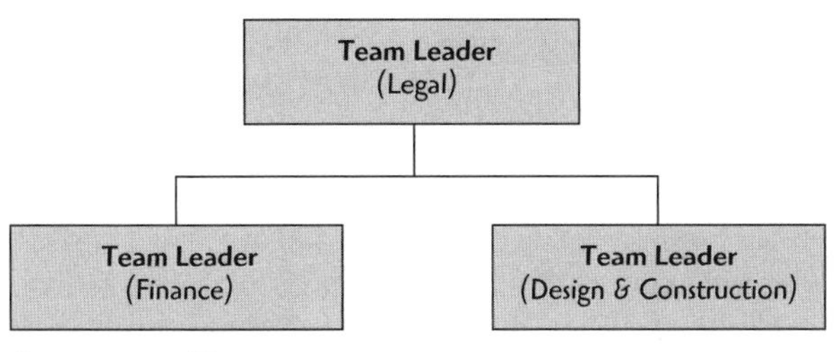

Figure 9.1 Procurement Team.

The following sections describe the specific *responsibilities* of the team members.

- *Team Leader (Legal) (Laverne Fleck).* Overall responsibility for all aspects of the procurement process, including the coordination of all activities related to solicitation, selection, and contract administration. The Team Leader (Legal) has primary responsibility to the following:
  - Manage the procurement team
  - Ensure that a competitive bidding process exists and that the steps outlined in this Procurement Plan are being followed
  - Provide direction and guidance to other team members
- *Project Leader (Design and Construction) (Jim Harris).* The major goal of the Project Leader on the Procurement Team is to prepare the technical criteria to be included in the Request for Proposal (RFP). Specific responsibilities are as follows:
  - Prepare expression of interests (EOI) as required for consulting firms
  - Prepare technical criteria for the RFP
  - Develop evaluation criteria (a sample of detailed evaluation criteria for architectural design services is included as Appendix A)
  - Review project proposals, supervision of the preliminary design and scale model, and detailed final design
  - Evaluate proposals
  - Provide oversight of the various consultants and contractors involved in the construction phase
  - Manage change control with respect to design and construction
- *Team Leader (Finance) (Steve Jackson).* Responsible for ensuring that competitive procurement practices are being followed and assisting the Project Manager with the financial issues related to the procurement. Specific responsibilities are as follows:
  - Ensure that there are at least three qualified bidders
  - Assist the Team Leader (Design and Construction) with development of evaluation criteria
  - Evaluate financial stability of firms
  - Establish financial conditions to be included in contracts
  - Evaluate proposals from a financial perspective

Procurement Strategy

KLSJ will act as overall *Project Manager* while contracting specific aspects of project execution to technical firms selected by KLSJ and approved by Carlington. To ensure that Carlington Aquatic Parks receives the best price for the work performed, the *overall procurement strategy* calls for all contracts to be bid competitively with a minimum of three qualified bidders. Contracts with an overall value estimated at more than Can$1 million must have a minimum of five qualified bidders. Any exception to the procurement strategy requires the approval of the Project Manager and Carlington Aquatic Parks.

The Team Leader (Design and Construction) will be responsible for the day-to-day management of the consulting and construction contracts. This will include the following:

- Providing direction to contractors and consultants
- Fielding questions from contractors and consultants
- Reviewing progress and certifying work as complete
- Authorizing changes in scope (see Table 9.1)

The Team Leader (Finance) will be responsible for signing off on all progress payments before any payments are issued to ensure that the funds have been allocated and approved and that the money is available for payment.

Scope of Activities

The *WBS* for the Ottawa Carleton Water Park outlined a number of tasks that will be procured from independent contractors. A complete list of the contracts to be procured is included in Table 9.1.

Procurement Schedule

Details of the procurement activities and their duration are contained in the Work Breakdown Structure. A detailed WBS and a procurement schedule for the procurement of detailed architectural design services have been included in Appendix B as a sample.

Risk Assessment

Recognizing and managing potential risks before they impact the project is an essential aspect of project management. *Risk management* is a proactive and iterative process; it is vital to controlling costs, meeting deadlines, and producing quality results.

KLSJ Consulting has conducted a thorough risk assessment of the contracts to be procured as part of the Water Park project. The criteria for evaluating risk are presented in Table 9.2. Complete details of the risk assessment are outlined in the *Risk Management Plan*.

Table 9.1: Procurement Contracts

Task	*Contract Type*	*Degree of Risk (to Project)*	*Estimated Costs*
Initial architectural design	Reimbursable costs with ceiling	Medium	Can$18,000
Market research	Fixed price plus incentive fee	Low	$12,000
Environmental study	Reimbursable costs with ceiling	Medium	$36,000
Archaeological study	Reimbursable costs with ceiling	Medium	$8,200
Traffic study	Reimbursable costs with ceiling	Medium	$15,200
Site services study	Reimbursable costs with ceiling	Medium	$26,600
Environmental assessment	Reimbursable costs with ceiling	Medium	$17,000
Detailed architectural design	Reimbursable costs with ceiling	Medium	$116,000
Design engineers	Reimbursable costs with ceiling	Medium	$262,000
Construction management firm	Fixed price with bonus and penalty clause for completion	Low	$10,000,000

Based on the risk assessment, KLSJ assesses that the consulting contracts are medium-risk, mainly because of the potential escalation of costs. The construction and market research contracts are considered low-risk.

A *Risk Management Working Group* (RMWG) has been established and will be responsible for all risk-monitoring and mitigation activities. The group will meet monthly to review the risk forms and react appropriately. All changes will be published. Special meetings will be arranged to address emergency items as required.

Although the initial risk management process has identified many potential risks associated with the project, new risks will continue to surface during the two and a half years of the project. The Risk Management Team will monitor the probability and impact of risks constantly and revise the project completion date, costs, and deliverable dates accordingly. Carlington Aquatic Parks will be consulted and advised on such changes if and when they occur.

Table 9.2: Risk Evaluation Criteria

Probability Criteria

Probability Rank	***Description***
High	More than 50% probability of occurring
Medium	Between 25 and 50% probability of occurring
Low	Less than 25% probability of occurring

Impact Criteria

Schedule Impact

Impact Rank	***Description***
High	Delay the opening of the park beyond July 1, 2005 (more than 6 weeks)*
Medium	Delay the opening of the park between June 15, 2005, and July 1, 2005 (3 to 6 weeks)
Low	Delay the opening of the park between May 24, 2005, and June 15, 2005 (less than 3 weeks)

Cost Impact

Impact Rank	***Description***
High	Could add more than 20% to cost of contract
Medium	Could add 10 to 20% to cost of contract
Low	Could add less than 10% to cost of contract

Project Review

KLSJ will monitor the ongoing progress of the project through regularly scheduled status meetings. Status meetings for the design phase of the project will occur biweekly; the frequency will increase to weekly during the execution phase (Phase 3) of the project. A representative from each of the consulting firms will be required to attend the weekly construction status meeting throughout the execution phase of the project.

Each consultant and contractor will be required to submit a copy of his or her project schedule to KLSJ Consulting at the start of the project and as requested throughout the project. Since KLSJ uses Microsoft Project as its project management software, firms will be requested to use it as well. If contractors fall behind on their schedules, they will be required to demonstrate to KLSJ Consulting how they plan to reallocate resources to get back on schedule. If appropriate, cost/time progress reports may be requested that use the Earned Value reporting method. The client and the vendor will agree on the format and content of these reports.

To a large extent, contract payments will be based on milestones, with representatives from KLSJ and Carlington Aquatic Parks approving the submission before authorization of payment.

Related Project Plans and Documentation

The project staff and contractors will require several *management documents* during the project. All project documents will be made available to the client and the relevant operations staff at specified milestones or at the end of the project. Documents may include project plans (including Work Breakdown Structures, schedules, cost plan, and resource plan), as well as appropriate technical product documents. Contractors will be provided with copies of studies and relevant information if it is available.

Terms and Conditions

The terms and conditions of the construction and consulting contracts will be based on standard documentation. All construction contracts will be based on Canadian Construction Documents Committee (CCDC) and Canadian Construction Association (CCA) standard construction documents, as published by the Canadian Construction Association (see Appendix C). In addition, all subcontractors must be bound by the same standard agreements that applies to the prime contractors.

Assumptions

The Procurement Plan described in this report is based on the assumption that Carlington Aquatic Parks is operating as a private organization. As such, KLSJ is in a position to invite firms to submit proposals for all aspects of the project. The contracts do not have to be open to all firms or advertised publicly. The Project Plan is founded on the following *key assumptions*:

- The project is being conducted by a private operation and is not subject to government and Treasury Board purchasing guidelines.
- The project will not be subject to any NAFTA or free trade regulations.
- KLSJ can invite firms to bid on the project.
- Contractors of sufficient size and expertise are available.

KLSJ believes that these assumptions are reasonable. If any of the assumptions prove otherwise, however, there undoubtedly will be an impact on project's cost, time, and/or quality. A risk analysis, as described previously, has been completed, and mitigation action as applicable will be undertaken to minimize these impacts.

Approvals

This Procurement Plan outlines the details of all procurement activities. However, unforeseen circumstances may require that actual procurement activities deviate from the Procurement Plan to maintain the project schedule, budget, or desired level of quality. Any deviation from the procurement strategy requires the written approval of the Project Manager and Carlington Aquatic Parks.

Approval from both the Team Leader (Design and Construction) and the Team Leader (Finance) is required to issue progress payments. The Team Leader (Design and Construction) will be responsible for certifying that the work has been completed to a satisfactory level, and the Team Leader (Finance) will certify that the funds have been allocated and approved and that the money is available for payment.

A member of the KLSJ management team must authorize all changes to established contracts. The Team Leader (Design and Construction) will review all proposed changes and, if warranted, will approve the change or recommend approval to a higher level of authority, as detailed in Table 9.3.

Table 9.3: Signing Authority Levels

Authority	*Authorization*	*Authority Level*
Team Leader (Finance)	• Certify progress payments	• Unlimited provided that money has been authorized and is available
Team Leader (Design and Construction)	• Certify progress payments • Changes within original scope of work	• Unlimited • Consulting: Can$5,000 • Construction: $25,000
Project Manager	• Deviations from the procurement plan • Changes within original scope of work	• Consulting: $10,000 • Construction: $50,000
Carlington Aquatic Parks	• Deviations from the procurement plan • Changes within original scope of work • All changes not included within original scope of work	• Unlimited • Unlimited • Unlimited

Summary

KLSJ Consulting has developed a comprehensive Procurement Plan that will allow it to oversee the conceptual development, design, approval, construction, testing, and handover of the new Ottawa–Carleton Water Park. The Procurement Plan will maximize the probability of completing the facility as designed, on time, and on budget.

As was the case with the original Project Plan, approval of this Procurement Plan by the managers of Carlington Aquatic Park is required within the next 14 days in order for KLSJ to maintain the Project Plan schedule.

References

1. "Few Waves Yet to Ride," *Link Magazine*, Spring/Summer 1999, p. 26.
2. Rakos, J., *Software Project Management for Small to Medium Sized Projects*, Prentice Hall, Englewood Cliffs, NJ, 1990, p. 152.

Appendix A: Architectural Design Detailed Evaluation Criteria

The evaluation criteria for architectural design are shown in Exhibit 9.1.

Exhibit 9.1 Form to evaluate architectural design.

PROPOSAL ITEM	WEIGHT	SCORE (1–10)	TOTAL (WEIGHT*SCORE)
1. Previous relevant experience (designed water park of at least Can$8 million in value)	Mandatory		
2. Project Team composition	10		
3. Previous projects completed on time and within budget	8		
4. Financial stability	8		
5. Use of technology	6		
6. Location of offices	4		
7. Deliverables	4		
Grand total			
Cost			
Cost/point			

Appendix B: WBS and Procurement Schedule for Detailed Architectural Design Services

The WBS and Procurement Schedule is shown in Figure 9.2.

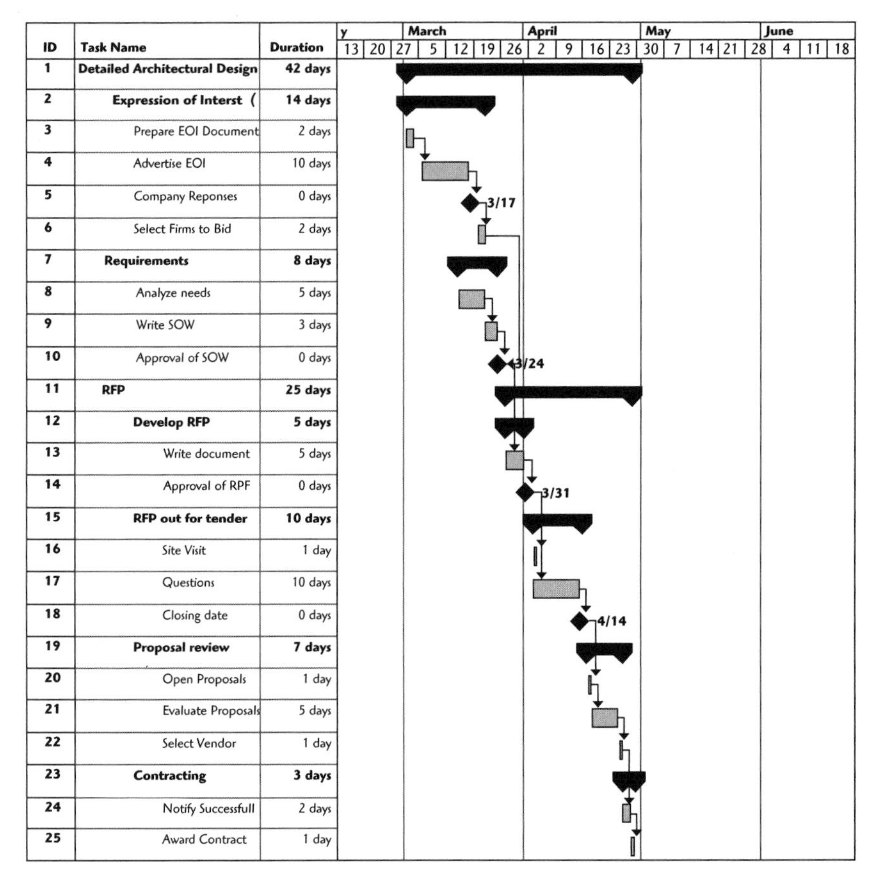

ID	Task Name	Duration
1	Detailed Architectural Design	42 days
2	Expression of Interst (	14 days
3	Prepare EOI Document	2 days
4	Advertise EOI	10 days
5	Company Reponses	0 days
6	Select Firms to Bid	2 days
7	Requirements	8 days
8	Analyze needs	5 days
9	Write SOW	3 days
10	Approval of SOW	0 days
11	RFP	25 days
12	Develop RFP	5 days
13	Write document	5 days
14	Approval of RPF	0 days
15	RFP out for tender	10 days
16	Site Visit	1 day
17	Questions	10 days
18	Closing date	0 days
19	Proposal review	7 days
20	Open Proposals	1 day
21	Evaluate Proposals	5 days
22	Select Vendor	1 day
23	Contracting	3 days
24	Notify Successfull	2 days
25	Award Contract	1 day

Figure 9.2 Architectural Design WBS and Procurement Schedule.

Appendix C: CCDC and CCA Standard Construction Contracts

Available from Canadian Construction Association. Use your local standard.

10

Acceptance Test Plan

Acceptance Test Plan: Discussion

The Project Team must deliver a product or service that meets all the functional specifications identified in the contract. The *acceptance process* provides written confirmation that the customer has accepted the product or service that was promised in the original contract. Formal acceptance is a key milestone. It signals that the end of the execution phase is near and usually triggers the final contract payments as well. The acceptance documents need to be clear and concise as they represent the written verification of a formal handover of responsibility to the customer. The acceptance process may even involve a ceremony or series of events attended by the customer and project team representatives. Success is achieved by demonstrating that the capability and functionality of the deliverables meet the promises and expectations.

The *acceptance plan* is a key component of the project plan. This plan defines the acceptance criteria, details the tests that will measure the performance of the product against the functional specification standards, and outlines the arrangements for the actual tests, including the schedule, location, participants, and testing method.

Testing for large projects may include only a representative sample of the final set of products. Performance standards normally are explained in the original contract documents for the project. The customer scores the acceptance tests against these agreed-on performance standards. Acceptance criteria normally include quantitative assessments of the following:

- Meeting design specifications for expected functionality
- Meeting design specifications for appearance
- Meeting design specifications for performance
- Operating in worst-case environmental conditions
- Demonstrating quality and reliability
- Demonstrating quality of the user documentation
- Demonstrating positive customer impressions

The promised training also must be accepted, although this may happen at a later date.

There are several *acceptance methods*. The simplest is the "trial period" approach, in which the client uses the product for a period of time. If it is satisfactory, it is accepted and paid for. There are two problems with this approach. First, at the end of the trial period the client may still feel apprehensive that he or she has not seen everything and that some problems may crop up in the future. Second, if even a small item is amiss, the client may demand that the clock for the trial period be restarted. In this case small problems may delay acceptance for a long time.

A better acceptance method is a thorough demonstration or test of every major function. This is, of course, slower and more costly than the trial period approach, but after such a demonstration, clients usually feel comfortable and are willing to pay.

Planning for acceptance is first done in the planning phase. The very first contract between the client and the vendor details how acceptance will be done. The acceptance plan itself, however, may be written only in the Execution phase, after the product design is complete. This is done because an organized demonstration of the product involves testing it piece by piece, and it is only after the design is complete that the pieces are known. The acceptance plan must be signed off by the client.

Acceptance Test Plan Outline

Executive Summary

This is an optional section that summarizes the major points of the document. It should provide an overview to readers who do not need to know all the details or so that they can decide if they need to read the remainder. At a minimum, it should identify the overall acceptance strategy that will be used on the project and highlight that this is key to completion and payment.

Introduction

Introduce the reader to the general purpose and method of acceptance. Emphasize the fact that if everything is accepted, payment must follow within a short period.

Participation

List who will be involved in the acceptance: who will be responsible for demonstrating capability, who will sign off on approvals, and the authority levels required. State what will be done if any item is not accepted.

Acceptance Criteria

List the conditions for acceptance; for example, all functions must meet all requirements completely, or only major functions will be tested and possibly less than 100% performance will be acceptable.

Appendix A: Acceptance Test Schedule

The test schedule usually consists of a table with the following columns:

Time	*Item*	*Project Team Test Coordinator*	*Client Representative*

Other details include the following:

- Where (test will be conducted)
- When (time of day and length of test)
- What (item to be tested)
- Who: project team test coordinator (responsible person for the contractor)
- Who: client representative (responsible person for the client)

Appendix B: Acceptance Test Method and Results

This usually consists of a table with the following columns:

Serial	*Test Subject*	*Test Parameters*	*Test Method*	*Score*	*Remarks*

Other details include the following:

- Where (test will be conducted)
- Serial (a unique test number)
- Test subject (category of the item to be tested)
- Test parameters (type of test, for example, environment, performance, appearance)
- Test method (detailed actions for the test)
- Score (the scale, such as a rating scale of 1 to 10)
- Remarks (any comments, such as a cause for nonacceptance)

Acceptance Test Plan: Example

KLSJ Consulting

January 1, 2003

Acceptance Test Plan

Ottawa–Carleton

Water Park

14 Palsen St., Ottawa, ON, Canada, K2G 2V8

Executive Overview

This document details how the client will accept the product. It details that there will be a formal demonstration of all the promised deliverables, and states who will participate, the schedule, and the criteria for acceptance. The table showing the items to be accepted, the criteria, and the acceptance results to be filled in is shown. If it is accepted, the client must pay 90% of the fees as per the contract.

General

KLSJ will conduct a formal acceptance test program for the Ottawa–Carleton Water Park. These tests will be reviewed by the representatives of Carlington Aquatic Parks, who are required to sign off on the acceptance test results prior to the completion of the project. The acceptance testing of the Water Park will be conducted during the week of May 9–13, 2005. A sample of each element of the park intended to be operational for the grand opening on May 21, 2005, will be acceptance tested. The testing schedule is shown in Appendix A. *If all items are accepted, 90% of the contracted amount will be paid to KLSJ, with a 10% holdback payable on completion of the warranty period.*

Participation

KLSJ will organize the preparations for the acceptance tests and provide a detailed schedule of activities no less than 90 days prior to the commencement of acceptance testing. Carlington Aquatic Parks will provide at least two representatives to witness each of the acceptance tests. These representatives will be asked to score and sign off on the Acceptance Testing Report. Therefore, we request that individuals with the appropriate knowledge and authority to sign be provided. Any aspects of the Water Park operation that fail the original acceptance test will be retested prior to the grand opening date. Failure to achieve project acceptance prior to the grand opening will result in contract penalties being applied to KLSJ, as described in the Master Contract Document.

Acceptance Criteria

All aspects of the Water Park are required to meet the performance and appearance standards in the Master Contract Document signed by KLSJ and Carlington Aquatic Parks in the fall of 2002. The Acceptance Test Results format is included in Appendix B.

Appendix A: Acceptance Test Schedule

All testing is to be conducted on May 9, 2005, at the Ottawa–Carleton Water Park site. See Table 10.1 for the test schedule.

Appendix B: Acceptance Test Method and Results

Table 10.2 shows the acceptance test method and results.

Table 10.1: Test Schedule

Time	*Item*	*KLSJ Test Coordinator*	*Client Representative*
May 6, 2005	Parking area	Test Coordinator 1	To be determined (TBD)
May 6, 2005	Grounds	Test Coordinator 2	TBD
May 6, 2005	Entrance	Test Coordinator 3	TBD
May 7, 2005	Concessions	Test Coordinator 1	TBD
May 7, 2005	Water slides	Test Coordinator 2	TBD
May 7, 2005	Wave pool	Test Coordinator 3	TBD
May 7, 2005	Restaurant	Project Manager	TBD
May 8–9, 2005	Attractions	Test Coordinator 1	TBD
May 8–9, 2005	Operations	Test Coordinator 2	TBD
May 10, 2005	Client selection	Project Manager	TBD

Grades:

A: Exceeds contract specification.

B: Meets contract specification.

C: Fails to meet contract specification owing to minor deficiencies. Requires correction.

D: Fails to meet contract specification owing to major deficiencies. Requires correction and retest.

Table 10.2: Scoring Test Method and Results

Serial	*Test Subject*	*Test Parameters*	*Test Method*	*Score*	*Remarks (for C or D)*
1	Slides	• Performance specifications	• Weighted dummy with instrumentation sent down each slide, minimum 10 times		
		• Environmental test	• Water quality test		
		• Appearance	• Visual inspection		
		• Overall impression			
2	Amusement attractions	• Performance specifications	• As specified in individual documents manufacturer		
		• Environmental test	• Water pumps and machinery in good working order		
		• Overall Impression	• Visual inspection		
3	Grounds	• Performance specifications	• Walk-around inspection; compare work to specifications and drawings		
		• Environmental test	• Visual inspection for spills; soil tests if warranted; confirm spills have been remediated		
		• Appearance	• Visual inspection		
		• Overall impression			
4	Safety systems	• Performance specifications	• Inspection against standards; test equipment at random		
		• Environmental test	• Inspect for leaks		
		• Appearance	• Visual inspection for obstructions		
		• Overall impression			

(Continued)

Table 10.2: Scoring Test Method and Results (*Continued*)

Serial	***Test Subject***	***Test Parameters***	***Test Method***	***Score***	***Remarks (for C or D)***
5	Food services and concessions	• Performance specifications	• As specified in individual manufacturer documents		
		• Environmental test	• Water quality tests; grease traps and ventilation testing;] public restroom testing		
		• Appearance • Overall impression	• Visual inspection		
6	Park operations staff	• Performance specifications	• Verify personnel documents for training in lifeguarding, water safety, cardiopulmonary resuscitation		
		• Environmental test	• Verify personnel docs for training in WHMIS, chemical storage, herbicide and pesticide use		
		• Appearance	• Observation during training		
		• Overall impression			
7	Transport	• Performance specifications • Environmental test • Appearance • Overall impression	• Visual inspection; test drive • Emissions test; visual inspection for leaks • Visual inspection		
8	Sanitation	• Performance specifications	• Test evacuation in liters per minute; test all pumps and drainage systems; test maximum pounds per square inch		
		• Environmental test	• Leakage inspection		
		• Appearance • Overall impression	• Verify systems not within view of public areas		

Reference 1: Amusement Slide Product Specifications. Park Equipment Corp., 1999.

Reference 2: Recommendations for Environmental and Safety Requirements for Outdoor Equipment. Health and Welfare Canada, Publication HW-3001, 1998.

Reference 3: Safety and Security Regulation in Public Places. Municipality of Ottawa, Publication OCR 201, 1998.

Reference 4: Specifications for Hygienic Food Services. Health and Welfare Canada, Publication HW-4201, 1998.

Execution Phase Documents

11

Project Meetings and Reports

Team Status Meeting: Discussion

Communications planning involves deciding who needs what information and when, where, and how it will be provided (see Chapter 6 for additional information on communications planning). An essential component of communications planning is the dissemination of information within the project team. Communication among the team members during project planning and execution is essential to maintain order, give the project manager and team leaders a sense of control over events, and maintain team rapport. A status meeting is a commonly used method to communicate general project information to the team and receive accurate and up-to-date work-in-progress information from team members.

The status meeting is essentially a review meeting used to inform team leaders and the project manager of project progress, resource usage, and other triggers that may have affected the schedule or budget since the last meeting. In a larger project, the various team leaders will chair their respective meetings; in a smaller project with only a few team members, the project manager most likely will chair it. In the former case, the team leaders then will ensure that the information is communicated to the project manager, whose responsibility it is to communicate with stakeholders external to the project team.

There are several additional objectives for this meeting, some of which are hidden:

- To allow team members to gather project information verbally from each other. This means that no one is in the dark about important new developments.
- To force people to report in front of their peers. This allows for accountability and maybe even some finger-pointing (this is a hidden objective).
- To allow for a group discussion of
  - Accomplishments to date (including schedule and budget performance)
  - Work packages
  - Action items
  - Technical concerns
  - Coordination issues

The frequency of status meetings depends on the scope of the project. Usually they are held once per week, but they also can be scheduled at specific times, such as at the start of a major phase, when certain milestones have been reached, or when a major problem has been encountered. It is best to hold the meeting at the end of the week rather than the beginning. If the meeting is held on a Monday, a significant amount of preparation for the meeting will be done on the weekend and the number of project hours is bound to increase dramatically. In contrast, if the meeting is held on a Friday, the preparation for the meeting is done during the week (and, miraculously, so is the rest of the work). Avoiding having people work over weekends and holidays goes a long way toward preventing project team burnout, not to mention improving team morale.

It is extremely important not to cancel or reschedule this meeting. Remain firm about meeting times, and people will learn to adjust their schedules accordingly. Keeping all the members of the team informed about project progress is critical to good project management. Not only does it eliminate miscommunication and misunderstandings, it brings the team members together by making them all part of the decision-making process.

To run a good team status meeting, several guidelines should be followed. The number of attendees should be minimized. Only the people who need to be there should attend, and they need to be warned well ahead of time. Although the meeting is relatively informal, there should be an agenda, with time allotted for each topic to be discussed. Stick to the agenda to avoid lengthy overruns to the meeting; that will be appreciated by all. Minutes or records of discussion with action items should be kept for future reference. The records of these meetings should be brief (a page or two at the most) and ideally promulgated the same day as the meeting. Action items must be followed up on, and an effective tracking system instituted. Note that records of status meetings provide an important input into the more formal project progress meeting with the client or project sponsor and the project status report that is discussed later in this chapter.

Team Status Meeting: Example

Team Minutes

CARLINGTON WATER PARK PROJECT

RECORD OF DISCUSSION WEEKLY TEAM STATUS MEETING MAY 16, 2003

Reporting period: May 5–May 16, 2003

Chairperson: Karen Dhanraj, Project Manager

Attendees: Scott Kennedy, Project Leader (Design and Construction)
Laverne Fleck, Team Leader (Legal)
Steve Jackson, Team Leader (Finance)

Absent: Jim Harris, Team Leader (Procurement and Risk Management)

1. **Activities and Accomplishments This Period**
   a. **Project Plan/Budget Approval.** The Project Manager (PM) reported that the Project Plan and project budget were completed and presented to Carlington Aquatic Parks on May 5, one week behind the baseline schedule. A subsequent meeting with the president of Carlington Water Parks on Thursday May 8 resulted in approval of the plan and budget as submitted. The PM noted that the client had expressed some concerns over cost estimates in several areas, as well as the risk assessment of the early part of the project.

      Action Required: None

   b. **Business Plan Progress.** Work continues on developing the Business Plan (50% complete), with a scheduled completion date of June 16. This work remains one week behind schedule but is not on the critical path.

      Action Required: Team Leader (Finance) will provide weekly updates on progress.

   c. **Documentation.** The Project Manager indicated that some project management documentation that was behind schedule has been completed and placed in the Project File.

      Action Required: None

2. **New Problems Encountered**
   a. **External.** The PM noted that the Carlington Water Park management expressed concerns with respect to the cost estimate of the water slides. This will require further investigation into these estimates.

      Action Required: PM will discuss with Jim Harris and have Procurement Team review the costs and the possibility of an alternative source of supply. Results of review to be reported in two weeks (May 26).

b. **Internal.** Scott Kennedy noted that a new project member would be joining his team in July to help prepare a series of presentations that will need to be made to various approval authorities in the fall. The individual will need office space.

   Action Required: Team Leader (Finance) will find suitable office space.

3. **Problems Solved**

   a. The Project Plan/Budget has now been approved as originally submitted.

4. **Problems Still Outstanding.** None

5. **Schedule Progress versus Plan and Trends**

   The PM reported that the project is now two weeks behind schedule. This is the result of the previous one-week delay plus the additional delay of one week during this period resulting from the delay in approval. It is intended to recover this slippage by sole-sourcing the environmental study to the consultant who conducted the initial environmental feasibility study. It is estimated that this will reduce this activity from four weeks to two, thus returning the project to its original schedule.

   A lengthy discussion regarding the schedule followed. Overall, all participants have confidence that the schedule can be adhered to.

   Action Required: Team Leader (Legal) will report at next status meeting progress on sole-sourcing the environmental study.

6. **Expenses versus Budget**

   The Team Leader (Finance) reported on the current budget status.

   a. Expenditures for this period were Can$9,750 for project management resources (labor).

   b. Expenditures to date of $177,000 are $5250 over budget.

   c. Reducing the duration of the environmental consultant hiring process is estimated to save $1000, for an overall estimate of $4250 (0.04%) over budget.

   Action Required: It was decided that the budget was still tracking within acceptable limits and that no extraordinary measures needed to be taken at this time. Team Leader (Finance) will continue weekly budget reports.

7. **Plan for Next Period**

   The PM reminded all involved that two new activities will commence in the next two weeks: selection of the environmental study consultant (critical path) and selection of the marketing company (non–critical path). Work on the Business Plan will continue throughout the period.

Project Progress Meeting: Discussion

The project progress meeting, sometimes called the project management meeting, is usually a meeting with managers to report progress. It can be a meeting of the Project Manager with the client, a meeting of the Project Manager with the Team Leaders on a project, or a meeting of the Project Managers in a company with upper-level management. The client may or may not be invited to some of these meetings. The project progress meeting is similar in many respects to the team status meeting; however, it is used to inform external stakeholders about the status of the project. The project progress meeting usually is held less often than the team status meeting, perhaps on a monthly or even quarterly basis. The frequency of the meetings, as well as specific reporting requirements, may be stipulated in the project contract.

The objectives for the progress meeting include the following:

- Monitoring of the project by outside stakeholders
- Reporting of progress to the stakeholders
- Reassuring the stakeholders (keeps them off your back)
- Warning the stakeholders early about problems (to avoid surprises)

From the project team, usually only the Project Manager will attend this meeting. For the client meeting, the project sponsor or other representatives may attend. Other stakeholders may be invited if issues that concern them will be discussed.

The normal practice is to keep formal minutes of the meeting, particularly if decisions are being requested from anyone. Normally, the Project Manager will present the latest status report, which will form the basis of discussion for the meeting. This may entail a formal presentation.

Project Progress Meeting: Example

Project Manager and Client Meeting Minutes

May 19, 2003

PROJECT MANAGER AND CLIENT MEETING HELD AT KLSJ CONSULTING HEADQUARTERS, 9:00 A.M., MAY 19, 2003

Attendees: Karen Dhanraj, Project Manager
Dan Milks, President, Carlington Aquatic Parks
Rod MacIvor, Representative, Environmental Engineering, Regional Government
Suzanne Blake, Administrator, Carlington Aquatic Parks

Old Business

1. *Project Plan/Budget Approval.* The cost of estimates and overall project risk were addressed. The Project Plan/Budget has been approved as originally submitted.
2. *Business Plan Completion.* The client was informed that work continues on developing the Business Plan (50% complete), with a scheduled completion date of June 16. This work remains one week behind schedule but is not on the critical path. The client is satisfied that the work will be done but expressed concern that any further delays could have an impact on the plan to obtain financing for the project.

Reports

3. *Schedule.* The project Status Report from May 16, 2003, was distributed. The Project Manager indicated to the client that the project is now two weeks behind schedule.

It is intended to recover this slippage by sole-sourcing the Environmental Study to the consultant who conducted the initial environmental feasibility study. It is estimated that this will reduce this activity from four weeks to two, thus returning the project to its original schedule.

Mr. MacIvor, the representative from the Environmental Engineering Department of the Region, stated that the Regional Government should approve both the consultant and the study before construction on the site is begun. Both the Project Manager and the client felt that this may add extra delays to the project. It was suggested that the credentials of the consultant be provided to Mr. MacIvor, to which he agreed. He pointed out that by law the Region must approve the Environmental Study and promised that he would be able to turn the study around within one week.

The client and the developer were both satisfied with this course of action.

4. *Budget.* The Project Manager informed the client that at this point the project was over budget by Can$4250, which was well within acceptable limits, and that no extraordinary measures were required at this stage. The client agreed.

New Business

5. The Project Manager reminded the client that work is progressing very well. Two new activities will commence in the next two weeks: selection of the environmental study consultant (critical path) and selection of the marketing company (non–critical path). Work on the Business Plan will continue throughout the period.

Adjournment

6. The meeting was adjourned at 10:30 a.m. The next project progress meeting will be held on June 16, 2003, at the KLSJ offices.

Project Manager and Steering Committee Meeting Minutes

May 20, 2003

PROJECT MANAGER AND SHAREHOLDER STEERING COMMITTEE MEETING HELD AT KLSJ CONSULTING HEADQUARTERS, 9:00 A.M., MAY 20, 2003

Attendees: Karen Dhanraj, Project Manager
Dan Milks, President, Carlington Aquatic Parks
Rene MacDonald, Chair, Board of Directors, Carlington Aquatic Parks
Marie Rakos, Board of Directors, Carlington Aquatic Parks
John Rakos, Board of Directors, Carlington Aquatic Parks

Old Business

1. *Project Plan/Budget Approval.* The cost of estimates and overall project risk were addressed. The Project Plan/Budget has been approved as originally submitted.
2. *Business Plan Completion.* The board was informed that work continues on developing the Business Plan (50% complete), with a scheduled completion date of June 16. This work remains one week behind schedule but is not on the critical path. The

client is satisfied that the work will be done, but expressed concern that any further delays could have an impact on the plan to obtain financing for the project.

Reports

3. *Schedule.* The project Status Report from May 16, 2003, was distributed. The Project Manager (PM) indicated to the board that the project is now two weeks behind schedule.
4. *Budget.* The Project Manager informed the board that at this point the project was over budget by Can$4250, which was well within acceptable limits, and that no extraordinary measures were required at this stage.
5. *Customer Satisfaction.* The Project Manager reminded the board that work is progressing very well and that the client is generally satisfied with the process.
6. *Risks.* The PM listed all the known risks:
   a. There is a risk of a two-week schedule slippage because the newspaper in which we were going to advertise the contract for the Environmental Study cannot publish it for two weeks. It is intended to recover this slippage by sole-sourcing the Environmental Study to the consultant who conducted the initial environmental feasibility study. It is estimated that this will reduce this activity from four weeks to two, thus returning the project to its original schedule.
   b. The board was informed that Mr. MacIvor, the representative from the Environmental Engineering Department of the Region, stated that the Regional Government may not be able to approve both the consultant and the study before construction on the site is begun. The PM related that a response strategy was developed: In the case of slow approval of the site, the issue will be raised to the Provincial Government.
7. *Issues.* The PM listed all the issues that arose from team meetings or other project meetings. There was concern about the performance of one team member, whose deliverables had a tendency to be late. The solution was to get his peers to remind him of the importance of delivering on time. This seems to have worked.

New Business

There were no new business items to discuss.

Adjournment

The meeting was adjourned at 10:30 a.m. The next project progress meeting will be held on June 20, 2003, at the KLSJ offices.

Project Status Report: Discussion

After the team status meetings, the Team Leaders meet with the Project Manager to report their individual teams' progress. The Project Manager then publishes a status report for the overall project. The status report is distributed internally as well as to the client and other stakeholders who wish to see it. The major purpose of this report is to facilitate monitoring of the project by these stakeholders. The status report not only contains the progress to date but, importantly, should provide early warning of potential problems. Taken together, the status reports provide a history of the project for new stakeholders and are the main source of input for the Post-project Report described in Chapter 16.

The Status Report should be published at least monthly, ideally to coincide with the project progress meeting, more often if people need to have timely information, and less often if there is not much progress during a specific period. The report should summarize the results of the team status meetings, the Project Manager's meetings, and any other relevant information for the reporting period.

Historically, the biggest problem with status reports (in fact, almost all reports) is that they are too long. A status report should be only a few pages long, with a couple of pages maximum for the text and a few pages of appendixes with a financial summary and a tracking Gantt Chart.

Project Status Report Outline

The contents of the status report should include the following sections.

Activities and Accomplishments for Reporting Period

This section is an explanation of any *new* accomplishments that have been completed since the last status report. Things completed are the most impressive. You should indicate whether there has been any change in the schedule since the last report and compare progress with the planned baseline. Refer the reader to the attached Gantt Charts. You also should state what has been started and how far along the new tasks are.

Problems Encountered

This section identifies problems encountered during the reporting period only. Explain not only the cause of an issue but, more important, the potential solutions. Address the expected impact on the budget, schedule, and quality as appropriate.

Problems Solved

Discuss any problems that were identified in the previous status reports and were solved in the reporting period. Document the solution and indicate its effect on the schedule and budget.

Problems Still Outstanding

Any problems identified in previous status reports that still have not been solved are listed *briefly* here. (One reason why status reports tend to be too long is that people rehash old problems. Spare everyone the time.) Succinctly point out any progress made and any new developments regarding the problem and its solution.

Time and Expenses

Resource and cost watchers will be interested in the expenditures over the period. A one-page financial summary attached as an appendix should suffice. Highlight in the body of the report significant differences from the planned budget (baseline), the likely impact on the final cost of the project, and if any mitigation measures are necessary. Client or Sponsor approval may be necessary if contingency funds are required.

Progress and Trend Analysis

Referring to the Gantt and financial charts, point out the calculated project completion date and estimated final cost. Trend analysis using techniques such as Earned Value, coupled with professional judgement, is invaluable in this regard. Chapter 10 in *The Guide to the Project Management Body of Knowledge* provides an introduction to the concepts of Earned Value; more detailed explanations can be found in most good project management textbooks. Project risk analysis may indicate that the project could get even further behind (or, rarely, ahead) of schedule, particularly if a high-risk phase of the project lies ahead. You owe the stakeholders the benefit of your insight and experience. Announce the possibility of delays or cost overruns; at worst you will have to admit later to the good news that you were wrong and the project did not lag further behind. Avoid the urge to be overly optimistic. The client will appreciate your honesty in the long run.

Plans for the Next Period

Briefly outline the work you expect to have completed in the following period as well as significant items that are to commence. Do not forget to highlight anticipated risk areas, if any, that are coming up. This helps reduce surprises in the next status report.

Project Status Report: Example

Water Park Project: Biweekly Status Report

Author: Karen Dhanraj, Project Manager

To: Carlington Aquatic Parks, Project Team, Project File

Reporting Period: April 28–May 16, 2003

Activities and Accomplishments This Period

The Project Plan and Project Budget were completed and presented to Carlington Aquatic Parks on May 5, one week behind the baseline schedule. However, approval of the Project Plan was not received until this morning (May 16), delaying the project by a further week. Work continues on developing the Business Plan (50% complete), with a scheduled completion date of June 16. This work remains one week behind schedule but is not critical path. Selection of the consultant for the environmental study and the marketing company, scheduled to commence this week, was postponed pending approval of the Project Plan and Budget.

Problems Encountered

The Carlington Water Park management expressed some concerns over the Project Plan/Budget with respect to several cost estimates, as well as risk assessment, which caused delay in the approval of the Plan/Budget.

Problems Solved

A meeting was held between the president of Carlington Water Parks and the Project Manager on Thursday, May 15, at which time concerns about cost estimates and risk were addressed. The Project Plan/Budget have been approved as originally submitted.

Problems Still Outstanding

The project began the period one week behind schedule and Can$5250 (0.05%) over budget. This was primarily because the assessment of site options took longer than expected and there were delays in initial design work and in obtaining preliminary NCC site approval. These delays were offset somewhat by earlier time gains in developing the Project Concept and the feasibility study. Minor cost overruns were experienced in developing the initial design and the scale model of the Water Park.

Schedule Progress versus Plan and Trends

As depicted in the Tracking Gantt Chart attached as Appendix A, the project is now two weeks behind schedule. This is the result of the previous one-week delay plus the additional slip of one week during this period resulting from the delay in approval. It is intended to recover this slippage by sole-sourcing the Environmental Study to the consultant who conducted the initial environmental feasibility study. It is estimated that this will reduce this activity from four weeks to two, thus returning the project to its original schedule. The forecast opening of the Water Park therefore remains May 21, 2005.

Time and Expenses

Expenditures for this period were Can$9750 for project management resources (labor). Expenditures to date of $177,000, as depicted in the Financial Report attached as Appendix B, are $5250 over budget. Reducing the duration of the environmental consultant hiring process is estimated to save $1000, for an overall estimate of $4250 (0.04%) over budget. The Earned Value Chart in Figure 11.1 also indicates that the project is tracking close to planned estimates.

Plan for Next Period

Two new activities will commence in the next two weeks: selection of the environmental study consultant (critical path) and selection of the marketing company (non–critical path). Work on the Business Plan will continue throughout the period.

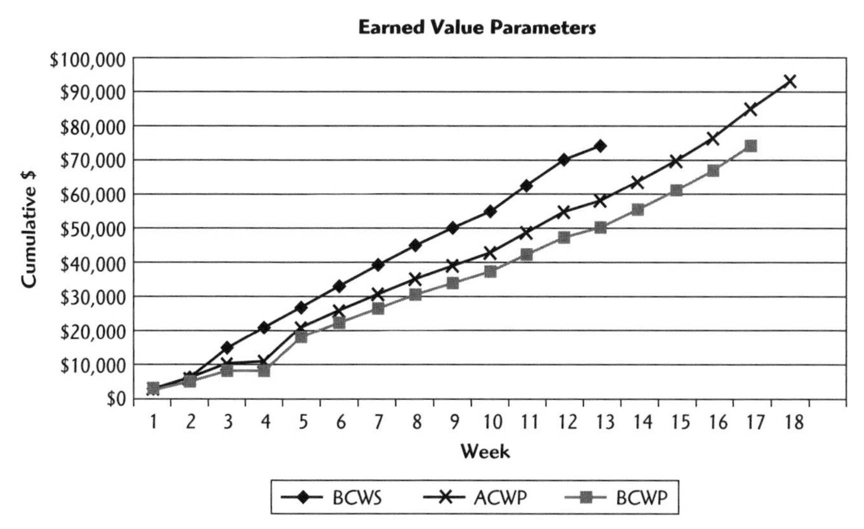

Figure 11.1 Earned Value parameters.

Appendix A: Tracking Gantt Chart

The Tracking Gantt Chart is shown in Figure 11.2.

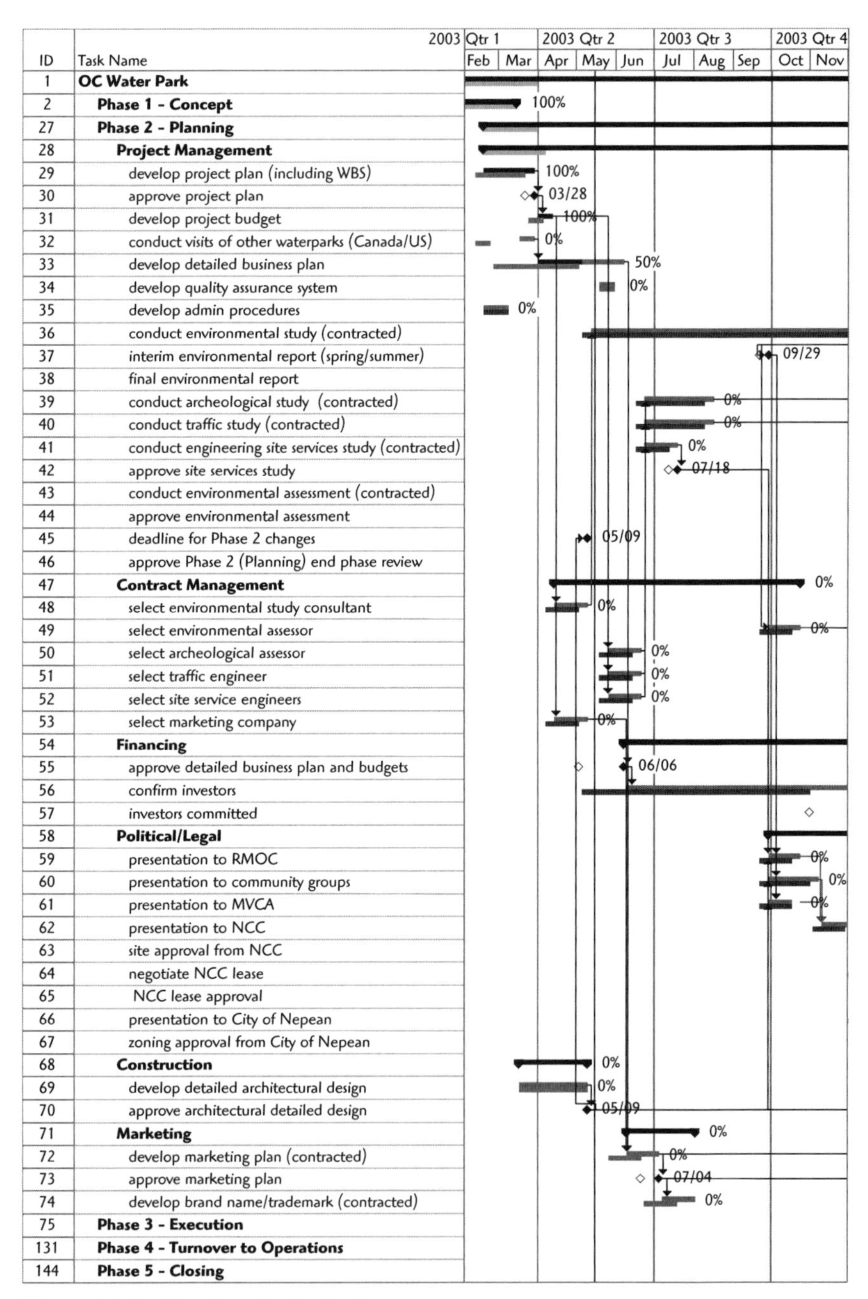

Figure 11.2 Tracking Gantt Chart.

Appendix B: Financial Report

The Financial Report is shown in Table 11.1.

Table 11.1: Financial Report (Can$)

Project Activity	*Planned Expenditures*	*Actual to Date*	*Remaining*	*Estimated Actual*	*Variance*
Phase 1: Concept	134,000	139,250	0	139,250	5,250
Phase 2: Planning					
Project Management					
Develop project plan	12,000	12,000	0	12,000	0
Develop project budget	4,500	4,500	0	4,500	0
Visit other water parks	10,000	10,000	0	10,000	0
Develop business plan	22,500	11,250	11,250	22,500	0
Develop quality assurance	1,600	0	1,600	1,600	0
Develop administrative procedures	3,750	0	3,750	3,750	0
Environmental study	36,000	0	36,000	36,000	0
Archaeological study	8,200	0	8,200	8,200	0
Traffic study	15,200	0	15,200	15,200	0
Engineering site services study	26,600	0	26,600	26,600	0
Environmental assessment	17,000	0	17,000	17,000	0
Subtotal	157,350	37,750	119,600	157,350	0
Contract Management					
Select environmental consultant	2,000	0	2,000	2,000	0
Select environmental assessor	2,000	0	1,000	1,000	(1,000)
Select archaeological assessor	1,600	0	1,600	1,600	0
Select traffic engineer	1,600	0	1,600	1,600	0
Select site service engineers	1,600	0	1,600	1,600	0
Select marketing company	2,000	0	2,000	2,000	0
Subtotal	10,800	0	9,800	9,800	(1,000)

(Continued)

Table 11.1: Financial Report (Can$) (*Continued*)

Project Activity	***Planned Expenditures***	***Actual to Date***	***Remaining***	***Estimated Actual***	***Variance***
Financing					
Confirm investors	52,000	0	52,000	52,000	0
Political/Legal					
Presentation to RMOC	3,600	0	3,600	3,600	0
Presentation to community groups	3,000	0	3,000	3,000	0
Presentation to MVCA	2,000	0	2,000	2,000	0
Presentation to NCC	7,400	0	7,400	7,400	0
Negotiate NCC lease	5,000	0	5,000	5,000	0
Presentation to City of Nepean	7,400	0	7,400	7,400	0
Subtotal	28,400	0	28,400	28,400	0
Construction					
Develop architectural design	116,000	0	116,000	116,000	0
Marketing					
Develop marketing plan	12,000	0	12,000	12,000	0
Develop brand name/trademark	1,000	0	1,000	1,000	0
Subtotal	13,000	0	13,000	13,000	0
Total Phase 2	377,550	37,750	338,800	376,550	(1,000)
Phase 3: Execution	10,718,100	0	10,718,100	10,718,100	0
Phase 4: Handover to Operations	98,500	0	98,500	98,500	0
Phase 5: Closing	33,000	0	33,000	33,000	0
Project Total	11,361,150	177,000	11,188,400	11,365,400	4,250

12

Risk Control Report

Risk Control Report: Discussion

As indicated in the Risk Management Plan (see Chapter 7), risk control is one of the four key steps in risk management and is essential to managing the outcome of a project. Risk management is an ongoing, iterative process. As the project progresses through its various stages, the probability of risks occurring and/or their level of impact on the project may change significantly. New potential risks may develop that were not foreseen at the start of the project. Regular monitoring and updating of risks as the project progresses is required to maintain an accurate assessment of overall risk for the project. Risk control outlines the procedures to be followed and the personnel involved in maintaining and updating the risk management log.

Risk control is an internal procedure that is intended primarily to provide the members of the project team with an understanding of the current state of risks on the project and the changing nature of the environment and outside influences. Given the dynamic nature of projects, the risk management log must be revisited and updated. This is done through regular meetings of the Risk Management Working Group (RMWG). Although the frequency of these meetings will depend on the size and scope of the project as well as the nature of the project environment, typically the risk management team will meet on a monthly basis. For larger, more complex projects and projects where the environment is particularly dynamic or unstable, meetings may occur more often. Depending on the severity, the RMWG may be called into an ad hoc meeting to react to a particular risk event. The RMWG typically will involve the same individuals who initially identified the project risks and created the associated mitigation strategies. As indicated in the Risk Management Plan, key stakeholders often are invited to be members of the RMWG. Their ongoing involvement in monthly risk meetings helps keep them informed about the changing nature of the project environment. In addition, their involvement helps the project team assess the political or intangible impacts associated with certain risks.

During the monthly meetings, the RMWG will identify any new risks that may have emerged during the project as well as revisit all the previously identified ongoing risks. All outstanding risks are reevaluated to determine their revised probability and potential level of impact on the schedule, cost, or quality of the project. In reevaluating risks, be sure to use the same criteria that originally were described in Risk Management Plan. If the need to modify these criteria arises, the new evaluation criteria should be defined clearly, and the new criteria should be applied to all outstanding and newly defined risks. The results of monthly meetings are documented in a risk control report. In addition, the current status of ongoing risks should be updated in the risk management log. Updating the risk table identified in the Risk Management Plan by using the recently updated risk assessments will provide a revised evaluation of the overall project risk classification. A well-documented risk report also can be an effective tool to disseminate information to stakeholders; risk control reports can form an integral part of the project's Communication Plan (see Chapter 6).

Good risk control also helps project managers anticipate and avoid potential risks before they result in serious problems. In reviewing and preparing a mitigation strategy, an effective risk control plan also will identify events or circumstances that should trigger a response so that managers can react early; you do not have to wait until a risk becomes a real problem before you react. The early implementation of mitigation strategies can assist in avoiding problems entirely or at least help minimize the impact of such an event.

If risk events do occur and become real problems, the risk control plan can help project managers react quickly and control "fires." Instead of the managers having to react in the heat of the moment, the mitigation strategies prepared earlier will provide a suggested course of action for each risk event. The mitigation strategy should be reviewed in light of the actual circumstances at the time; nonetheless, it provides an effective starting point in developing a response to unforeseen events.

There are instances when, regardless of how effective the risk planning and control procedures are, unforeseen circumstances occur and potential risks may result in problems or changes to the project. In these instances, contingency funds or management reserves can be used to allow for changes in the project requirements (see Chapter 14).

Risk Control Report Outline

This risk report generally takes the form of a memo or a record of discussion from the Risk Manager to the Project Manager and the members of the RMWG.

Activities and Accomplishments

Events that have been completed successfully are noted, and the risks associated with these activities can be closed. No further action on these items is required.

Plan for the Coming Month

Identify major events that are planned for the coming month and determine which risks may be affected as a result of the planned activity. Assess the overall environment of the project and consider external events that may affect the project.

Active Risks

Update existing risks, including a revised estimate of the impact on time, cost, and quality that may be associated with a risk event. Also reevaluate the probability of a risk occurring.

Open Risks

New risks that may result from the changing environment or unforeseen circumstances are identified and documented in the risk log, as detailed in the Risk Management Plan (see Chapter 7). Determine the potential impact of new risks to the project and develop a mitigation strategy.

Project Impacts

Report on the cumulative impact of risk on the project as a whole. Update the risk management summary table and revise the total project risk classification as required. Communicate any significant changes in the overall risk level to stakeholders.

Risk Control Report: Example

KLSJ Consulting

March 15, 2005

Risk Control Report

Ottawa–Carleton

Water Park

14 Palsen St., Ottawa, ON, Canada, K2G 2V8

Monthly Risk Report: March 15, 2005

From: Jim Harris, Risk Manager

To: Risk Management Working Group

Karen Dhanraj, Project Manager
Dan Milks, Carlington Aquatic Parks
Scott Kennedy, Team Leader (Design and Construction)
Laverne Fleck, Team Leader (Legal)
Steve Jackson, Team Leader (Financial)

Activities and Accomplishments February: March 2005

Initial advertising and promotion has begun, and several corporate sponsors have been identified and contacted. Procurement of tickets and passes is under way. No delays in these activities are foreseen. They are expected to be completed on their scheduled finish dates.

Plan for March—April 2005

Preparation for the trial opening will begin in one week. Immediately after that, the trial opening will take place from April 11, 2005 to April 22,

2005. Handover to the park operations team is scheduled for April 25, 2005.

Active Risks

The water slides were not delivered as of September 20, 2004. The manufacturer was contacted on October 25, 2004, and agreed to deliver the slides by March 22, 2005. The slides arrived on March 15, 2005. Installation is to begin immediately. The manufacturer has agreed to cover all costs involved with this late installation date. This risk item will remain open until the installation of the slides has been approved.

The park operations team still has not been hired. Since this task is not a critical path item, it has not resulted in schedule slippage yet. However, the handover to the park operations team, scheduled for April 25, 2005, is a critical path task. The team must be chosen prior to this date. At this time, the Project Manager (PM) is negotiating with two candidates for the position of Operations Manager. The issue involves payment terms for the manager. The risk item will remain open until the chosen manager signs his or her contract.

Open Risks

With the late installation of the water slides, there is a risk that the post-trial modifications and repairs will take more than the three weeks allocated in the schedule, causing the grand opening to be delayed beyond the May 21 long weekend. To mitigate this risk in advance, the majority of testing and balancing of the equipment will be started immediately and will continue during preparations for the trial opening. Therefore, only a few modifications should be required post-trial.

There is still a minor risk that financial difficulties may prevent the construction company from installing the water slides at this late date. There is also a minor risk that inclement weather may delay completion of the slide installation. These risks will remain open until final approval of construction is obtained.

Project Impact

The risks encountered have been mitigated, and most of them have been closed or are expected to close in the near future. The project remains on schedule, and it is expected that the May 21 opening date will be achieved.

Total project risk classification: low

The Project Risk Table is in Appendix A. The Risk Management Summary Table is in Appendix B.

Risk Log

The updated risk log for the various risks is shown in Exhibits 12.1 to 12.11.

Exhibit 12.1 Log for risk no. 1.

Water Park Risk Management Form	
Risk No: 1	**OPI**: Team Leader/(Finance)
Title: Funding is not secured in time to meet project deadlines.	
Description: Approximately Can$12 million in debt and equity financing is required prior to the beginning of Phase 3 (Execution). With an uncertain investment climate, it is probable that it will not be possible to secure the necessary funding we need in time to progress to construction in the summer of 2003.	
Probability: High	
May 1, 2003	*High*
June 19, 2003	*Medium*
December 15, 2003	*Low*
Level of Impact:	
May 1, 2003	*High*
June 19, 2003	*Medium*
December 15, 2003	*Low*
Overall	High
May 1, 2003	*High*
June 19, 2003	*Medium*
December 15, 2003	*Low*
Schedule	High
May 1, 2003	*High*
June 19, 2003	*Medium*
December 15, 2003	*Low*
Cost	Low
May 1, 2003	*Low*
June 19, 2003	*Low*
December 15, 2003	*Low*
Mitigation Strategy: Emphasize importance of investment during presentations to interested parties and Ottawa-area organizations. If investors are not forthcoming, approach NCC and City of Nepean for assistance. If financing not secured by the time site approval and zoning approval are awarded, halt the project and conduct a complete assessment of options. Do not proceed to Phase 3 without funding secured.	
Risk Monitoring: **May 15, 2003: Active** *At this time very few investors have been identified. During upcoming presentations to stakeholders and through radio and newspaper publicity, the need for funding will be emphasized. The client and Team Leader (TL) (Finance) will address these issues.* **June 19, 2003: Mitigated** *Potential investors have been identified but must be confirmed. TL (Finance) has been assigned to contact all investors and confirm their commitment to the project.* **December 15, 2003: Closed** *All investors have been confirmed. Risk item closed.*	

Exhibit 12.2 Log for risk no. 2.

Water Park Risk Management Form

Risk No: 2	**OPI**: Project Manger

Title: A political action group successfully petitions the NCC, resulting in delays or denial of site approval and subsequent delays in the start of Phase 3 (Execution).

Description:

There are a number of active organizations that oppose the construction of the Water Park in its current location. These groups, acting alone or together, could direct sufficient political attention toward this issue to force the NCC into protracted community consultations. Community hearings and studies could delay the issuance of site approval by many months, making on-time completion of the project impossible.

Probability: High	
November 27, 2003	*High*
December 4, 2003	*High*
December 22, 2003	*High*
January 5, 2004	*High*
January 19, 2004	*Low*
Level of Impact:	
Overall	High
November 27, 2003	*High*
December 4, 2003	*High*
December 22, 2003	*High*
January 5, 2004	*High*
January 19, 2004	*Low*
Schedule	High
November 27, 2003	*High*
December 4, 2003	*High*
December 22, 2003	*High*
January 5, 2004	*High*
January 19, 2004	*Low*
Cost	Low
November 27, 2003	*Low*
December 4, 2003	*Medium*
December 22, 2003	*Medium*
January 5, 2004	*Low*
January 19, 2004	*Low*

(continued)

Exhibit 12.2 *Continued*

Mitigation Strategy:

Hold preemptive meetings with all concerned parties to hear their points of view. Develop and implement a comprehensive communication plan to counter all valid arguments against the park.

Have a second, more aggressive plan to counter specific groups that actively lobby the NCC. If delay appears imminent, hire a government relations firm to assist with obtaining political approval.

If delay actually occurs, suspend all project activity and reassess options for completion by original deadline.

Risk Monitoring:

Date/Status:

November 27, 2003: Active

Presentations to community groups were given. Since then, several action groups have petitioned the NCC to deny site approval for the Water Park. The PM and Project Leader (PL) (Design) have been assigned to host a general meeting to reassure all involved parties about their concerns.

December 4, 2003: Mitigated

The preemptive information meeting was held. The political action groups were not satisfied with the arguments given for construction of the Water Park. They indicated that they would continue to pressure the NCC to deny site approval. The PM has been assigned to hire a lobby group to counter the arguments presented by the political action groups.

December 22, 2003: Mitigated

The scheduled presentation to the NCC was given today. Site approval was not given. The PM and PL (Design) have been assigned to give another presentation to the NCC on January 4, 2004. The lobby group will continue to pressure the NCC for site approval.

December 6, 2004: Mitigated

NCC site approval was granted. However, the NCC lease approval and subsequent zoning approval from the City of Nepean have been delayed. This may delay Phase 2 approval. TL (Legal) has been requested to speed up lease negotiations with the NCC.

December 17, 2004: Closed

NCC lease approval has been granted. The project is on schedule once again. Risk item closed.

Exhibit 12.3 Log for risk no. 3.

Water Park Risk Management Form	
Risk No: 3	**OPI**: Project Manager
Title: Unfavorable environmental assessment demands major mitigation.	
Description: The result of the environmental assessment reveals information that may have a major impact on the viability of the project. Environmental concerns may result in significant conditions being placed on approval for the land and zoning approval from the NCC and from the City of Nepean. These conditions may range from minor solutions to major and potentially costly remediation. The level of time, cost, and effort required to mitigate these environmental concerns may prove to be so demanding that the viability of the site becomes questionable.	
Probability: Medium	
May 8, 2003	*High*
June 12, 2003	*High*
July 17, 2003	*High*
July 24, 2003	*High*
September 25, 2003	*Low*
Level of Impact:	
Overall	High
May 8, 2003	*High*
June 12, 2003	*High*
July 17, 2003	*High*
July 24, 2003	*High*
September 25, 2003	*Low*
Schedule	High
May 8, 2003	*High*
June 12, 2003	*High*
July 17, 2003	*High*
July 24, 2003	*High*
September 25, 2003	Low
Cost	High
May 8, 2003	*High*
June 12, 2003	*High*
July 17, 2003	*High*
July 24, 2003	*High*
September 25, 2003	Low

(continued)

Exhibit 12.3 *Continued*

Mitigation Strategy:

Conduct an initial review of the property to determine previous owners and users. Conduct an initial environmental assessment to determine potential obvious risks prior to committing much time, effort, and money to the process. If these initial studies do not reveal any significant risks, proceed with a detailed environmental assessment to be completed by an independent contractor. Environmental damage resulting from the prior operation of the sewage treatment plant should be identified clearly and documented. The responsibility for the environmental cleanup of these prior contaminants should remain the responsibility of the NCC.

Risk Monitoring:

Date/Status:

May 8, 2003: Active

An initial environmental assessment was performed. It indicates possible environmental damage to Watt's Creek and the surrounding fauna due to prior operation of the sewage treatment plant. The NCC will need to be informed of the damage. Cleanup and full environmental assessment must be completed before February 2004 if project is to remain on schedule. PM has been assigned to present results to the NCC and request cleanup.

June 12, 2003: Mitigated

The NCC has been approached, and cleanup of the site has been requested. PM has been assigned to follow up on site cleanup.

July 17, 2003: Mitigated

Cleanup has still not begun on Water Park site. This item is still high-risk, as final environmental report is required by January 5, 2004. PM has been assigned to transmit urgency of request to the NCC.

July 24, 2003: Mitigated

NCC has begun cleanup of the Water park site. Completion date is expected to be September 25, 2003. PM has been assigned to monitor this activity.

September 25, 2003: Closed

Cleanup of the site has been completed. Project is once again on schedule. Risk item closed.

Exhibit 12.4 Log for risk no. 4.

Water Park Risk Management Form	
Risk No: 4	**OPI**: Team Leader (Design and Construction)
Title: Construction schedule delayed due to delays with the delivery of attractions.	
Description: Many of the attractions for the Water Park, such as the water slides, are highly specialized items. Most water park attractions are custom designed and fabricated for each individual park. There are very few firms that fabricate water park attractions, and since each attraction is custom designed and built, the producer has limited capacity. Any delays in production, not only of our attractions but of attractions for other parks that may come before ours in the production cycle, may have a negative impact on the delivery of our attractions.	
Probability: Low	
September 20, 2004	*High*
October 18, 2004	*High*
October 25, 2004	*High*
March 15, 2005	*Medium*
Level of Impact:	
Overall	Low
September 20, 2004	*High*
October 18, 2004	*High*
October 25, 2004	*High*
March 15, 2005	*Medium*
Schedule	Low
September 20, 2004	*High*
October 18, 2004	*High*
October 25, 2004	*High*
March 15, 2005	*Medium*
Cost	Low
September 20, 2004	*Low*
October 18, 2004	*Low*
October 25, 2004	*Low*
March 15, 2005	*Low*

(continued)

Exhibit 12.4 *Continued*

Mitigation Strategy:

When selecting firms to fabricate the Water Park attractions, distribute the work evenly among a few producers. If one firm fails to deliver, at least the park will have some attractions delivered on schedule.

Ensure that there are minimal design difficulties with the attractions to avoid unnecessary delays. Establish a milestone on the Work Breakdown Structure (WBS) to ensure that the attractions (e.g., slides) are ordered at the earliest possible date. The attractions are to be completed and installed prior to the end of the December construction. Contracts with the attraction manufacturers should contain a penalty clause stating that if attractions cannot be installed in the fall, as scheduled, the manufacturer will be responsible for additional costs to have the attraction completed in April/May, prior to the grand opening.

Risk Monitoring:

Date/Status:

September 20, 2004: Active

Water slides have not been delivered. This will delay their installation prior to the end of the December construction period. TL (Design) has been assigned to contact the water slide manufacturer to expedite delivery.

October 18, 2004: Mitigated

Water slides have still not been delivered. The manufacturer has cited overcapacity as the reason for this delay. TL (Design) will contact manufacturer to confirm delivery prior to preparation for trial opening on March 22, 2005.

October 25, 2004: Mitigated

The water slide manufacturer has been contacted. An agreement has been made for delivery by March 22, 2005. The manufacturer will assume all additional costs for installation at this late date.

March 15, 2005: Mitigated

Water slides have been delivered. Installation will begin immediately. TL (Design) has been assigned to monitor installation to ensure the grand opening date of May 18, 2005.

Exhibit 12.5 Log for risk no. 5.

Water Park Risk Management Form

Risk No: 5	**OPI**: Project Manager
Title: Inability to hire suitable operations manager at a reasonable salary.	

Description:

The ongoing success of the Water Park depends on the ability of the operations manager. Prior experience operating a similar facility is critical to the success of the park. There are very few water parks in Canada, and as a result, there are few qualified candidates. There are more qualified operators in the United States, but many of them are reluctant to move to Canada. Those who are prepared to relocate demand high salaries.

Probability: Medium	
August 16, 2004	*High*
January 7, 2005	*High*
March 15, 2005	*Medium*
Level of Impact:	
Overall	Medium
August 16, 2004	*High*
January 7, 2005	*Medium*
March 15, 2005	*Medium*
Schedule	Medium
August 16, 2004	*High*
January 7, 2005	*Medium*
March 15, 2005	*Low*
Cost	Low
August 16, 2004	*Low*
January 7, 2005	*High*
March 15, 2005	*Medium*

Mitigation Strategy:

Start recruiting for a qualified operations manager early in the process, at least one year prior to the opening date.

Promote the opportunities and rewards of running a new water park and being able to design the systems and controls. Promote the quality of life aspect of living in Canada and in the Ottawa region.

Offer bonus incentives based on operating profit and possible stock options in lieu of an excessive salary. There will be greater motivation for the operations manager to succeed, plus it will encourage the manager to remain in the position for a longer period.

Risk Monitoring:

Date/Status:

August 16, 2004: Active

After several weeks of recruiting efforts for a suitable operations manager, it has become apparent there is little interest in management of the Water Park. PM has been assigned to begin promotional efforts to attract new candidates.

(continued)

Exhibit 12.4 *Continued*

January 7, 2005: Mitigated

Several management candidates have been identified; however, they are unwilling to accept the position without a significant pay increase. PM has been assigned to negotiate an alternative payment method.

March 15, 2005: Mitigated

Two final candidates have been identified. They are both interested in alternative payment methods, including performance bonuses. The PM has been assigned to negotiate with both candidates and choose one.

Exhibit 12.6 Log for risk no. 6.

Water Park Risk Management Form

Risk No: 6	**OPI**: Team Leader (Design and Construction)

Title: Final consultant design not acceptable to the owners and investors.

Description:

One or more major aspects of the detailed design are not acceptable to the owners and/or investors when submitted for final approval. There are significant changes to the overall plan to accommodate site conditions and technical obstacles. The design does not meet the owner's expectations or changes the operations and/or image of the park in a significant manner.

Probability: Low	
July 7, 2003	*Low*
Level of Impact:	
Overall	Low
July 7, 2003	*Low*
Schedule	Low
July 7, 2003	*Low*
Cost	Low
July 7, 2003	*Low*

Mitigation Strategy:

Design consultant will provide monthly updates to the management committee and inform it of any changes to the concept as a result of the design process. The Team Leader (Design and Construction) will maintain close contact with the designers to ensure that the owner's interests are represented, including participation in biweekly design meetings.

The design consulting company's contract should establish payment milestones based on completion and acceptance of design documents.

Risk Monitoring:

Date/Status:

July 7, 2003: Closed

The detailed design was accepted and approved. Risk item closed.

Exhibit 12.7 Log for risk no. 7.

Water Park Risk Management Form

Risk No: 7	**OPI**: Team Leader (Design and Construction)

Title: Project may be delayed as a result of financial difficulties with construction company.

Description:

The Water Park will enter into a contract with the construction company. If the construction company experiences financial difficulties, payments to subcontractors and suppliers will cease, resulting in construction liens being placed on the property and the facilities. Removing the original contractor and replacing it would be extremely difficult.

Probability: Low	
March 15, 2005	*Low*
Level of Impact:	
Overall	High
March 15, 2005	*High*
Schedule	High
March 15, 2005	*High*
Cost	Medium
March 15, 2005	*Medium*

Mitigation Strategy:

When selecting a construction firm, ensure that Project Management Team conducts a thorough review of the financial history of the firms invited to bid by using services such as Dun & Bradstreet. The construction tender package should include the requirement that the successful firm be bonded. Both a performance and a labor and material bond from a certified bonding company should be required.

Risk Monitoring:

Date/Status:

March 15, 2005: Open

This risk item will remain open until all construction has been completed and approved.

Exhibit 12.8 Log for risk no. 8.

Water Park Risk Management Form

Risk No: 8 | **OPI**: Team Leader (Finance)

Title: A major investor withdraws from the project after construction has started.

Description:

Once construction has begun and the attractions have been received, the project is financially committed for substantially the entire Can$12 million-plus project cost. The withdrawal of a major investor at this point would make project completion tenuous at best and also could jeopardize scheduled payments to our firm.

Probability: Low	
March 15, 2005	*Low*
Level of Impact:	
Overall	High
March 15, 2005	*Low*
Schedule	High
March 15, 2005	*Low*
Cost	Low
March 15, 2005	*Low*

Mitigation Strategy:

Insert penalty clauses and withholding fees so that investors are discouraged from withdrawing after the commencement of Phase 3. Once construction has begun, put a builder's lien on all structures to ensure preferential treatment in the event of bankruptcy. If a major investor still withdraws, cease all construction and freeze all project accounts pending completion of a full financial reassessment.

Risk Monitoring:

Date/Status:

March 15, 2005: Closed

The final slides are being installed. The grand opening is in two months. There is no risk of cessation of construction due to investor withdrawal. Risk item closed.

Exhibit 12.9 Log for risk no. 9.

Water Park Risk Management Form

Risk No: 9	**OPI**: Team Leader (Design and Construction)

Title: Construction permits are not approved in time to complete construction before winter.

Description:

Construction cannot commence until all necessary permits have been obtained. These permits are subject to a number of approval levels within the city, any of which could request additional information or substantiation. Depending on the nature of the request, resubmission could delay receipt of permits by several weeks, making it impossible to complete construction before winter weather makes construction impractical.

Probability: Low	
July 19, 2004	*Low*
Level of Impact:	
Overall	Low
July 19, 2004	*Low*
Schedule	Low
July 19, 2004	*Low*
Cost	Low
July 19, 2004	*Low*

Mitigation Strategy:

Brief officials at the city on the project and solicit their input into the design process at an early stage. The design consultant should develop a relationship with city officials during the life of the project.

Submit draft plans to applicable city offices for comment during Phase 2 and correct as necessary. If plans are still rejected, reassign all available design resources to this task until the situation is corrected. If delays persist, halt all preparatory construction work and completely reassess all critical time lines to determine if completion date is still viable.

Risk Monitoring:

Date/Status:

***July 19, 2004*: Closed**

The construction permits have been approved. Risk item closed.

Exhibit 12.10 Log for risk no. 10.

Water Park Risk Management Form

Risk No: 10	**OPI**: Team Leader (Design and Construction)

Title: Inclement weather causes construction delays.

Description:

Unseasonal weather or inclement weather causes delays in the construction process and prevents major construction activities (such as pouring concrete) from proceeding as scheduled. Typical construction schedules would allow for delays on the basis of average days of work lost based on past seasonal experience, but extended unforeseen periods of inclement weather would not be taken into account.

Probability: Low	
March 15, 2005	*Low*
Level of Impact:	
Overall	Low
March 15, 2005	*Low*
Schedule	Low
March 15, 2005	*Low*
Cost	Low
March 15, 2005	*Low*

Mitigation Strategy:

The construction company may be able to manipulate some of the scheduled activities to suit the weather forecast. Overtime should be considered when forecasts of inclement weather threaten crucial critical path activities.

If the construction schedule slips, construction should be scheduled on weekends and possibly holidays to make up for time lost due to inclement weather.

Risk Monitoring:

Date/Status:

March 15, 2005: Open

This risk item will remain open until all construction has been completed and approved.

Exhibit 12.11 Log for risk no. 11.

Water Park Risk Management Form

Risk No: 11	**OPI**: Team Leader (Design and Construction)

Title: Post-trial modifications and repairs take more than the three weeks allocated in the schedule, causing the grand opening to be delayed beyond the May 21 long weekend.

Description:

There is a three-week window between the trial opening and evaluation and the grand opening celebrations. During this period modifications and repairs will be carried out on noted deficiencies.

Major unforeseen circumstances could require that significant modifications be made to the operating setup.

Major modifications could take longer than three weeks to identify, design, and implement.

Probability:	Medium
March 15, 2005	*Low*
Level of Impact:	
Overall	Low
March 15, 2005	*Low*
Schedule	Low
March 15, 2005	*Low*
Cost	Low
March 15, 2005	*Low*

Mitigation Strategy:

Involve the operators early in the process. The majority of testing and balancing of the mechanical equipment can be completed prior to the trial opening and evaluations. The operators should be present during the commissioning stage of the project and should sign off that the equipment is indeed in good working order.

Ensure that the operators have a detailed plan identified as a milestone on their payment schedules. This plan should be reviewed and approved by the owners prior to acceptance.

Risk Monitoring:

Date/Status:

March 15, 2005: Mitigated

Testing and balancing are to begin immediately for equipment already constructed and/or installed. This will prevent delay of the grand opening if testing and balancing take longer than the scheduled amount of time.

Appendix A: Project Risk Table

The project risk table is shown in Table 12.1.

Total project risk classification: low

Table 12.1: Project Risk, Revised March 15, 2005

Probability ***Impact***	***LOW***	***MEDIUM***	***HIGH***
HIGH	7. Project delayed due to financial difficulties of construction company.		
MEDIUM		4. Delayed delivery of attractions leads to construction delays 5. Inability to hire suitable operations manager.	
LOW	1. Funding not secured. 2. Political group petitions NCC. 3. Unfavorable environmental assessment. 6. Final design not acceptable to owners. 8. Major investor withdraws after project begun. 9. Construction permits not approved in time to finish before winter. 10. Inclement weather causes construction delays.	11. Post-trial modifications take longer than allocated time.	

Appendix B: Risk Management Summary Table

The risk management summary table is shown in Table 12.2.

Table 12.2: Water Park Risk Management Summary, Revised March 15, 2005

Risk	*OPI*	*Probability*	*Impact*	*Status*
1. Funding not secured	TL (Finance)*	High	High	Closed
2. Political groups delay approval	PM	High	High	Closed
3. Unfavorable environmental assessment	PM	Medium	High	Closed
4. Delay in delivery of attractions	TL (Design and Construction)	Low	Low	Mitigated
5. Unable to hire suitable operations manager	PM	Medium	Medium	Mitigated
6. Final design not acceptable	TL (Design and Construction)	Low	Low	Closed
7. Financial difficulties with construction company	TL (Design and Construction)	Low	High	Open
8. Major investor withdraws	TL (Finance)	Low	High	Closed
9. Delays associated with construction permits	TL (Design and Construction)	Low	Low	Closed
10. Inclement weather delays construction	TL (Design and Construction)	Low	Low	Open
11. Post trial modifications exceed three-week window	TL (Design and Construction)	Medium	Low	Mitigated

*TL: Team Leader; PM: Project Manager.

13

Quality Assurance and Quality Control Reports

Quality Assurance and Quality Control Report: Discussion

The focus of this chapter is on reporting on project quality to ensure that quality is properly controlled throughout the project and that the customer is ultimately satisfied with the quality of the end product. However, before getting too far into a discussion of project quality control, it is useful once again to remind the reader of the three principal project variables: cost, quality, and time. It always must be recognized that maintaining quality typically comes at a price in money, time, or both. Because of this, quality reporting from a control and assurance perspective is a very important aspect of any project.

Quality control deals first of all with controlling the quality of the product or service the customer eventually will receive. It also involves controlling the processes that make that product or service possible. For instance, destructive testing can verify the strength of concrete samples used in a building, but control of the concrete mixing and curing *process* is what offers the greatest guarantee of consistent quality. In addition, expensive testing and sampling often can be avoided through superior process control. The management of the project itself ensures quality control of the project management processes.

There are four reasons why firms engage in project quality control. First, they want to create a system that prevents failures from occurring. Second, they want to monitor the process and product periodically to ensure that failures are not occurring. Third, if a failure does occur, they want to detect it internally, that is, before the customer sees it. Fourth, if a failure occurs after the customer receives the good or service, they want to find out why it happened, correct it, and feed the information back into the control process. Studies have shown consistently that failure prevention is by far the least expensive means of delivering quality products or services.

There are two aspects to project quality control: the detection of faults, if any, and the correction of the causes of those faults, which results in continuous improvement. Both aspects should be based on the organization, functions, processes, and activities described in the Quality Management Plan (see Chapter 8). Any real-time changes that need to be incorporated into the quality control system should be fed back into the plan through a formal revision process.

A multitude of statistical tools and techniques can be employed to control the quality of the product and the process. Everything from basic flowcharts and run charts to sophisticated statistical process control (SPC) and six-sigma quality programs can be used. In fact, the use of statistics to control quality is the subject of numerous textbooks and is outside the scope of this chapter. Nonetheless, the reader should seek out an expert in this area because it is difficult to go beyond rudimentary control procedures without some use of statistics. The ISO 9001:2000 standard described in Chapter 8, for example, requires that a firm engage in statistical process control. Any firm seeking ISO registration will need to devote resources to this area. Project control methods documented in the *Guide to the Project Management Body of Knowledge* (for example, cost and schedule reporting using Earned Value) is one way to control the management of the project.

One can hope that all projects will be executed flawlessly and that no problems will ever occur. The reality, however, is that quality problems will occur and the project team needs to be prepared to deal with them constructively. Once a quality problem is discovered, there are some definite actions that are required, including the following:

- Actions that deal with the problem itself to bring the process back under control, correct the product defect, or both (the most likely problems should have previously documented actions to take to correct these errors)
- Actions to report the problem, the known or likely causes, the corrective action that was taken, the degree of success of that action, and any impact on the project in terms of cost, quality, and/or time

There are two basic choices for reporting the results of quality control monitoring. It can be done on a periodic basis, or it can be done on an ad hoc exceptions-only basis. Both methods have their merits, and it would be unwise to choose one method and stick to it doggedly. For instance, some clients may want to be reassured weekly that quality is under control, whereas others may want to know only when something out of the ordinary happens. When a problem does occur, some stakeholders may need to be informed immediately, whereas others can wait until the next regularly scheduled report (by which time the problem should have been resolved). A judicious combination of both routine and ad hoc reports will provide sufficient assurance without burying the project team in report writing.

The two types of quality reports most often used in a project are a quality assurance report and a quality control report. A quality assurance report consists of a few pages on how the quality management plan is being followed. It usually is done monthly or less frequently and may be done by exception or on demand. A quality control report is a summary (sometimes with supporting detail) of actual process and product measurement results, typically based on well-defined specifications set out by the client. It is done more frequently, based on a regular (possibly published) schedule, and definitely will be presented at all progress meetings. Examples of these two reports are shown later in this chapter.

Note: Regardless of the immediate usefulness of any quality measurement data, the data should all be archived for future reference as a permanent project quality record (see Chapter 16).

Quality Assurance Report Outline

Executive Summary

This is an optional section that summarizes the major points of the document. It gives readers a chance to decide whether they need to read the remainder. This section would be done only by exception, since most quality assurance reports are short enough not to need an executive summary.

Background

Keep the length of this section to a minimum. Simply refer to the major aspects of the project on which quality depends and explain why this led to an emphasis on certain quality control aspects in the quality management plan. If there is a particular reason why this quality assurance report is being produced, that should be mentioned.

Quality Assurance Focus

This portion should be based on the product features, specifications, and key processes identified in the quality management plan, other client-furnished documents, or published regulations. Tell the reader what major areas are being monitored and why and be sure to state the "why" in terms of consequences (favorable or unfavorable) to the client or other stakeholders. Obviously, most quality monitoring is done to prevent a negative consequence, but it is also worthwhile to measure things that ensure a positive consequence. A good mixture of both will demonstrate to the client that there is a thorough quality assurance program in place.

Quality Assurance Implementation

Based on the factors described above, describe the main areas where monitoring and testing are taking place. Here it may be necessary to be fairly specific, and so appendixes should be a consideration for truly large projects. Otherwise, a table will generally suffice to show what the desired result is, how it is being measured, the frequency and/or quantity of measurement, and how many measurements have been taken so far. If measurement has not yet begun, insert the date when it is scheduled to begin according to the project Work Breakdown Structure.

Quality Assurance Results

Here you should report clearly and concisely how effective the measurement and testing have been at ensuring that the quality standards are being met. Ideally, there will be nothing but good news. However, do not shy away from stating that errors have occurred (if this is the case) as long as you also describe what has been done to improve the offending processes or procedures. The point of the report is to assure the reader that quality standards are being met even if that means pointing out shortcomings and corrective actions. If quality control is being taken seriously by the Project Team, it will always be possible to report bad news in a positive way.

Conclusions

Do not make the readers draw their own conclusions. State the team's conclusions in the most objective and balanced way possible. Statements should be unambiguous, with an emphasis on facts and deductive reasoning, not intuition. If the current methods are not working, say so and explain why. If everything is working fine, try to be humble.

Recommendations

Normally, action has already been taken and problems have been corrected. However, sometimes an observation leads to the conclusion that the standard or feature is unachievable or that it will require more time and/or resources than are available. If this is the case, the client's consent will be needed through a formal change process (scope change or specification change). A recommendation to this effect (with details and substantiation) will get the change process started so that other team members will take appropriate action. For more details on change control, see Chapter 14.

Quality Control Report Outline

Executive Summary

This is an optional section that summarizes the major points of the document. It gives readers a chance to decide whether they need to read the remainder of the document. For a quality control report, the need for an executive summary hinges on the length of the main text and the level of technical sophistication needed to understand it. All long documents should be summarized, but short ones written in highly technical jargon (by engineers or statisticians, for instance) should have a summary at the front in plain language. It is also easier to transfer the executive summary into minutes of progress meetings and similar correspondence without having to reinvent the wheel.

Background

Once again, keep the length of this section to a minimum. Refer to the major aspects of the project on which quality depend, and explain why this led to an emphasis on certain quality control aspects in the Quality Management Plan. Comments should be limited to the areas for which quality measurement or testing are being done. If there is a particular reason why this quality control report is being produced (for example, as a response to a quality problem), that should be mentioned also.

Quality Control Focus

This portion should be based on the product features, specifications, and key processes identified in the Quality Management Plan, other client-furnished documents, or published regulations. Tell the reader what major areas are being measured or tested and why, and be sure to state the "why" in terms of consequences (favorable or unfavorable) to the client or other stakeholders. Obviously, most quality monitoring is done to prevent a negative consequence, but it is also desirable to measure things that ensure a positive consequence. A good mixture of both will demonstrate to the client that there is a thorough quality control program in place.

Note that whereas the quality assurance report can discuss less statistical outcomes (especially if they relate to providing assurance to the client), comments in the quality control report are limited to the areas actually being measured, verified, and tested. The quality control report also should report on the quality of the project process itself by reporting progress using any or all of the following: Earned Value (EV), Cost Variance (CV), Schedule Variance (SV), Cost Performance Index (CPI), and Schedule Performance Index (SPI). For more information on status reporting, see Chapter 11.

Quality Control Methodology

Describe the main areas where monitoring and testing are taking place. It may be necessary to be fairly specific here, and so appendixes should be included for large projects. Otherwise, a table will generally suffice to show what the desired result is, how it is being measured, the frequency and/or quantity of measurement, and how many total measurements have been taken so far. If measurement has not yet begun, insert the date when it is scheduled to begin according to the project Work Breakdown Structure.

Quality control can be a fairly complex activity involving very technically sophisticated models and formulas. If you must explain what is being done, be careful to consider who will be reading the report. Avoid jargon and technical terms as much as possible; it is unlikely that your client will be impressed.

Quality Control Measurements

Here you should briefly describe the measurements that have been taken, which can be summarized in a number of ways. Simplicity is once again the key. Mention high or low values, any out-of-limit values, any test failures, and similar figures. Percentages and standards often permit the reader to put things into perspective quickly. Do not regurgitate pages of data that only confirm that nothing is wrong and never mention a data point without putting it in context.

Quality Control Analysis

This section can be separate or can be rolled into the measurement section described above. Here you should report clearly and concisely what the measurement and testing activities have uncovered. Stay at the highest level possible without distorting the truth. For instance, instead of listing all 10 items being measured in process X and then stating that each of these items is operating within limits, simply state that there are no problems to report in process X.

Ideally, there will be nothing but good news. However, do not shy away from stating that errors or omissions have occurred if this is the case. Describe what has been done to investigate and correct the processes or procedures. Negative results should lead to status meetings, performance reports, and other remedial activity that in turn will lead to timely project decisions. The point of the report is to inform the reader whether the quality standards are being met and, if they are not, to advise on actions being taken or approvals needed.

Conclusions

The analysis from this section should lead to certain conclusions, either positive or negative, with respect to project control. Once again, do not shy away from negative conclusions if they are necessary. Objectivity will be rewarded if you have a solid plan for remedying the situation

Recommendations

Action on critical items probably has been taken by the time the report has been written. These decisions should still be written into the report. Most likely some of the recommendations will lead to a formal change process (scope change, cost, quality, and time change). If this is the case, state the requested change clearly so that the client understands that a decision is required.

Quality Assurance Report: Example

KLSJ Consulting

May 11, 2004

Quality Assurance Report

Ottawa–Carleton

Water Park

14 Palsen St., Ottawa, ON, Canada, K2G 2V8

Executive Summary

The attached report on quality assurance identifies one relatively minor incident involving a noncompliant subcontractor that has been resolved with no immediate effect on the schedule or budget. There are no other pending or outstanding issues to report.

The subcontractor incident pointed out the need for increased vigilance during the peak construction period that is about to begin. After reviewing the incident, the Project Manager (PM) has concluded that two additional quality auditors will be needed for approximately six months at a total cost of Can$60,000. The PM requests this change be approved as a formal change order (paperwork to follow), with the source of funds being the project contingency fund. The contingency fund is still well within tolerable limits.

This is a formal project report, and the recommendations contained herein require approval by the client.

Contents

Quality Assurance Implementation
Quality Assurance Results
Conclusions
Recommendations

Background

As stated in the Quality Management Plan, safety during construction and after the Water Park project has been completed are two major focus areas for quality assurance. To achieve the highest possible degree of confidence in the completed work, KLSJ is placing a great deal of emphasis on compliance with standards, building codes, and design specifications. In all, over 3000 individual areas are being measured or monitored to provide quality assurance to the client.

Quality Assurance Focus

KLSJ is using the ISO 9001:2000 standard for all aspects of the Water Park project. However, the focus is on a few key areas to give maximum quality assurance to the client. These areas are the following:

- All KLSJ documents and all subcontractor documents containing project standards and specifications are reviewed prior to use to ensure that they comply with stated standards and specifications. Project Manager sign-off is required prior to publication.
- All KLSJ documents and all subcontractor documents containing project standards and specifications are color-coded and serial-numbered. Copying is not permitted. All obsolete documents are collected and archived or destroyed to avoid any chance of faulty design or construction.
- Subcontractor processes and completed work are inspected daily by a KLSJ contract manager to ensure compliance to standards.
- All product arriving at the construction site is inspected prior to use.
- All nonconforming product is quarantined in a special lock-up with restricted access and removed from the work site as soon as is practical.
- All incidents and accidents (human or material) are documented immediately and reported to one of the on-site managers; the severity is assessed, and appropriate measures are taken as soon as possible.
- All significant accidents or incidents are reviewed at a weekly internal project meeting to determine if a broader quality assurance response is necessary and to keep all managers alert for similar occurrences in their areas of the site.

Table 13.1: Quality Assurance Areas Being Measured or Monitored

Result Expected	***Measurement Description***	***Frequency and/or Quantity***	***Total Measures Taken***
Compliance with building codes	Design control	As amendments are made	11
	Document control	Continuous	N/A
	Inspection of material	As it arrives; continuous	N/A
	Process audits	Weekly or more frequently if needed	57 audits on 14 separate major work processes
	Inspection of finished work	As work is completed; continuous	N/A

Quality Assurance Implementation

As was mentioned above, the KLSJ quality assurance plan encompasses over 3000 individual items. Table 13.1 highlights the general areas being measured or monitored.* More detailed information on each of these areas is available on request.

Quality Assurance Results

Since the last report, there has been only one relatively minor incident involving a noncompliant subcontractor who failed to produce the required City of Ottawa certification. This subcontractor has been removed from the site. A subsequent 100% recheck of all subcontractor certifications was conducted without incident. Inspection of certification documents has been added to the checklist that must be completed prior to receiving a work site security pass.

Since then, all process audit results have been positive. There are no other outstanding issues to report.

Conclusions

At this early point in the project quality is well under control. The subcontractor infraction was discovered during a routine process audit, and immediate action was taken to (1) remove the person from the work site, (2) have all work performed by this individual inspected by a certified technician, and (3) amend work procedures to preclude similar incidents.

*Only one example is given. The actual table would be much longer and may be better treated as an appendix.

Although this incident is considered minor, it emphasizes the need to maintain a high audit standard with frequent checks of key processes. The Project Manager has concluded that during the six-month period of peak construction activity, two additional audit teams will be needed on a part-time basis to maintain quality assurance and control.

Recommendations

It is recommended that a formal change order be initiated with the client to increase the number of audit teams by two during peak construction. Total additional cost is estimated at Can$60,000. This new requirement is to be paid from the contingency fund, which is well within tolerance limits at the moment.

Quality Control Report: Example

KLSJ Consulting

August 10, 2004

Quality Control Report

Ottawa–Carleton

Water Park

14 Palsen St., Ottawa, ON, Canada, K2G 2V8

Executive Summary

The attached report on quality control identifies one incident involving substandard lumber and one serious fluctuation in the newly installed power grid. Both problems have been resolved with the suppliers, with no adverse effect on the schedule or budget.

Two other potential areas of difficulty also have been noted. First, the large preexisting building is undergoing structural assessment, with a preliminary report due soon. Any serious problems identified in this report probably will result in an increase in both the time and the cost of the project. The second issue relates to the risk associated with key suppliers, with KLSJ proposing to conduct an internal evaluation and put in place a mitigation strategy where necessary.

This is a formal project report, and the recommendations contained herein require approval by the client.

Contents

Quality Control Methodology
Quality Control Measurements
Quality Control Analysis
Conclusions
Recommendations

Background

In the Quality Management Plan, control of the design and construction processes, particularly specifications and standards, was identified as the major focus area for quality control. To achieve the highest possible degree of confidence in the completed work, KLSJ is placing a great deal of emphasis on compliance with these standards, building codes, and design specifications. In all, over 3000 individual areas are being measured or monitored to provide quality control for this project.

Quality Control Focus

KLSJ is focusing on a number of key areas to guarantee maximum quality control to the client. Within each area, a number of subprocess measurements are being taken. To keep this report concise, only major processes are discussed unless a detailed explanation is necessary. These major areas are the following:

- Design verification and validation to ensure compliance with standards, specifications, codes, and other client-defined needs
- Evaluation of suppliers, subcontractors, purchased products, and material to ensure that standards of workmanship and quality are met
- Disposition of nonconforming product prior to use to ensure that substandard product is not used on the work site
- Testing and evaluation of completed work to ensure conformance with standards, specifications, codes, and other client-defined needs
- Documentation and archiving of quality records to make it possible to demonstrate to the client, stakeholders, and regulators that quality standards are being met

Quality Control Methodology

Table 13.2 highlights the general areas being measured or monitored.*
More detailed information on each of these areas is available on request.

*Only one example is given. The actual table would be much longer and may be better treated as an appendix.

Table 13.2: Quality Control Areas Being Measured or Monitored

Result Expected	*Measurement Description*	*Frequency and/or Quantity*	*Total Measures Taken*
Compliance with building codes.	Design control	As amendments are made	11
	Document control	Continuous	N/A
	Inspection of material	As it arrives; continuous	N/A
	Process audits	Weekly or more frequently if needed	57 audits on 14 separate major work processes
	Inspection of finished work	As work is completed; continuous	N/A

Quality Control Measurements

Design changes are defined as those which are significant enough to require a change in project scope. To date, there have been 11 major changes to the design of the Water Park. Each change was done according to the published document control process, and subsequent audits have revealed that no unauthorized or out-of-date specifications are being used.

Inspection of incoming products and material has led to relatively low levels of rejected product since the last report. Two weeks ago, the lumber supplier was changed because of an intolerable quantity of substandard materiel, all of which had to be quarantined and returned. The new supplier is completely satisfactory so far.

Destructive testing of concrete samples has revealed no abnormalities. Initial testing of the electrical grid detected two potentially serious fluctuations, both have which have been redone and tested to 100% compliance. Structural testing of the preexisting building is ongoing, with results expected in time for the next report.

Quality Control Analysis

Quality control issues have been dealt with as they were discovered, and all are now well within tolerable limits. In many areas, there has yet to be a single reported quality issue despite ongoing rigorous quality audits. However, the problem with the lumber supplier led to minor slowdowns in a number of areas, which pointed out the vulnerability of the project to slippage by key suppliers.

Conclusions

In general, the required level of quality control is being achieved and is even exceeding expectations in many areas. The problem with the supplier has pointed out a project vulnerability that is being looked into for possible weakness in other suppliers or subcontractors.

Recommendations

It is recommended that all critical component contracts be reviewed and that the risk and probability of nonperformance be assessed in detail. Any contracts that are deemed to be high-risk must have a specific, approved mitigation strategy in place as soon as possible. The criteria for identifying high-risk contracts will be developed separately in consultation with the Project Risk Manager.

14

Problem Report and Change Request

Problem Report and Change Request: Discussion

Problem and Issue Management

Project integration control involves controlling aspects of the project that affect some or all of the other eight knowledge areas in the *PMBOK Guide*. Part of integration control is the detection of problems and the reaction of a project team to those problems. Problems may result in changes to the plans, and this requires change management. Documents that are used include a problem report, a change request, and responses to those reports.

When problems arise during a project, they should be addressed immediately. They should be communicated to everyone involved and documented for the future. The status report (see Chapter 11) or an issue log can be used to document the cause of a problem, the responsible person, and the solution. Ideally, all parties should negotiate and come to an agreement on the impact of any change on the budget and schedule before that change is implemented formally. This can save countless headaches one is when trying to explain a cost rise or schedule slip to the stakeholders after the work has been completed.

Change Management

In the case of any change request, the Project Manager (PM) must "push back" by inquiring about the need for the change and determining whether it somehow can be avoided. Most important, the PM must research the cause of the change request. It may be something that the PM can fix. For example, if an external resource is not providing the promised deliverables, the PM must get on the phone or e-mail immediately and do everything in his or her power to rectify the situation.

Either the client or the contractor may request a change. A change can affect scope, time, or cost; however, a change in one usually affects the other two. Scope change frequently involves adding more work, a cost change requests additional funding, and a time change requests additional time, although changes can remove, modify, or eliminate work from a project as well.

Scope Change

Scope changes may result from new requirements for the project, a redesign, a technology change, or a business change. Scope changes must be submitted to the PM on a change order form (see below). To calculate the level of impact, it is important to look at the Work Breakdown Structure (WBS) and see what impact the change has on it. Usually a scope change results in additions to the WBS, but occasionally some tasks can be eliminated. Check the dependencies, as they often change as well. Then calculate the overall effect on the project time. Extra work requires time to complete. All these changes invariably affect the total cost. Fill this in in the level of impact section of the change order form and insist that the client authorize the change (and impact) before proceeding. The reaction is documented in the comments and rationale section of the form, and if any part of the project parameters is affected, all the stakeholders must be informed.

Cost Change

The project costs can increase when the resources used are more expensive than expected, overtime is required, equipment costs rise owing to external financial factors, or initial cost estimates are poor to begin with. Cost changes must be submitted to the PM on a change order form (see below). If the maximum project cost is fixed, cost savings must be realized some other way. The project may be rescoped (reduce the functions) or slowed down (sometimes the cost is less if resources can work off hours, such as weekends, but be aware of morale problems arising from too much overtime). Wise managers have a contingency or management reserve fund available, which allows a cost rise without too many problems. Fill this in in the level of impact section of the change order form and insist that the client authorize the change (and impact) before proceeding. The reaction is documented in the comments and rationale section of the form, and if any of the project parameters is affected, all the stakeholders must be informed.

Time Change

You may need additional time to complete the project because of poor time estimates, revised priorities (such as a key resource being removed from the project), delays in the delivery of resources, or other changes that affect time (scope and cost). Time changes must be submitted to the PM on a change order form (see below). To calculate the level of impact, always update the schedule with all known past and future changes. The additional time required may not affect the critical path. Look at what may be affected later. For instance, a one-month delay in one task (for example, design) may create a six-month delay later in the project (building).

If the project time is slipping, you can try to catch up by fast tracking: Squeeze the critical path by overlapping the end of one task with the start of another. This may require additional resources. You also can crash the schedule. This usually involves speeding up critical items with additional resources, overtime, improved technology, and so forth. Note that these responses all entail additional money.

Since it is extremely difficult to catch up once the schedule has slipped, a simple approach, when possible, is to replan the time. Update the schedule and announce the slip. This is very common in project management because it is the least risky solution: You are working with the same plan, just a little delayed. The reaction is documented in the comments and rationale section of the form, and if any of the project parameters is affected, all the stakeholders must be informed.

Problem Report and Change Request: Example

Problem/Issue Report No. 4

Friday, May 9, 2003

From: Karen Dhanraj, Project Manager

To: Dan Milks (Carlington Aquatic Parks), Jim Harris (Team Leader, Design and Construction), Laverne Fleck (Team Leader, Legal) and Steve Jackson (Team Leader, Financial)

Problem: Delay in approval of the Project Plan and Budget

Responsible: Client

As of today the project is two weeks behind schedule. Previous slippage has been compounded by the additional delay in approval of the Project Plan and Budget. If there is a further delay in approval of the Project Plan and Budget, it will delay the grand opening of the Water Park. According to the present schedule, the grand opening will be on June 1, 2005, at the earliest. This delay has resulted in the project being Can$5250 over budget. If the project remains behind schedule, it will be at least $5250 over budget.

Proposed Solution

If the plan and budget are approved by next Monday, the Project Team can recover this slippage by sole-sourcing the Environmental Study to the consultant who conducted the initial Environmental Feasibility Study. It is estimated that this will reduce the contracting process from four weeks to two, thus returning the project to its original schedule. This is also expected to decrease costs since the contracting process will take much less time.

Update: June 2, 2003

The Project Team successfully sole-sourced the Environmental Study to the consultant responsible for the Environmental Feasibility Study. This has brought the project back on schedule and has saved $1000 in sourcing costs.

Change Order Form: Example

A sample change order form is shown in Exhibit 14.1.

Exhibit 14.1 Change order form.

KLSJ Consulting

Project: Ottawa–Carleton Water Park		**Change No**: 1	
Submitted by: Karen Dhanraj		**Date**: May 12, 2003	

Title: Reduce Time for Procurement of Environmental Study

Description of the change:

Reduce the time required to contract for the Environmental Study. The project is currently two weeks behind schedule. Sole-sourcing the Environmental Study to the consultant who completed the Environmental Feasibility Study will save two weeks of contracting time and $1000.

Level of Impact:		Quality	Nil
Current Contract Value:	$2000		
Value of This Change:	($1000)	Current completion date	June 2, 2003
Revised Contract Value:	$1000	Revised completion date	June 20, 2003

Comments/Rationale:

Low-risk solution to return the project schedule to the original dates.

Approved:

KLSJ Consulting, Karen Dhanraj: ____________________

Carlington Aquatic Parks, Dan Milks: ____________________

15

Acceptance Report

Acceptance Report: Discussion

The *acceptance* is done toward the end of the Execution phase. If it is planned well, acceptance is simply a matter of running through all the tests and demonstrations defined in the Acceptance Test Plan (see Chapter 10). Preparations for the acceptance include the following:

- A specific time and place is set; all those who will attend are advised.
- All documentation is present, especially the signature forms.
- Knowledgeable technical people are present in case something has to be fixed quickly or questions arise.

The project team already has run through the acceptance process; if the documents are disorganized and ill-prepared, the client will be apprehensive and reluctant to pay you.

The acceptance report contains the results of the acceptance tests. This report is used to obtain formal acceptance of the deliverables by the client. The acceptance report describes and summarizes the results of the acceptance tests. If any problem areas are identified, the proposed solutions, including retesting and improvements or repairs to the product, are described. The acceptance report is usually a mandatory deliverable because it is the predecessor to the key milestone of final contract approval and payment.

The format of the acceptance report is flexible, but at a minimum it should include the acceptance test results and a discussion of any areas of concern along with proposed solutions. The actual content of the report obviously depends on the output of the project.

Acceptance Report Outline

Introduction

Introduce the reader to the purpose of thorough demonstration. Then go on to say when and where the acceptance was performed.

Test Results

This is simply the table as outlined in the acceptance test plan (included here or as shown in Appendix A). The columns "Score" and "Remarks" are filled in.

Acceptance

This is the key signature form. The signature of the client indicates acceptance. Complete payment (with a possible holdback until the end of the warranty, depending on the contract) must follow.

Acceptance Report: Example

KLSJ Consulting

May 17, 2005

Acceptance Report

Ottawa–Carleton

Water Park

14 Palsen St., Ottawa, ON, Canada, K2G 2V8

Introduction

Acceptance is the activity that demonstrates to the client that all the promised deliverables actually have been accomplished and that all the functionality works as promised.

The acceptance testing of the Ottawa–Carleton Water Park was conducted during the week of May 9–13, 2005. The Project Manager, the Design Team Leader, and the three Design Team members represented the KLSJ Project Team. The President of Carlington Aquatic Parks, the Chief Legal Officer, and the Operations Manager represented the client.

Test Results

The results of the testing are shown in Appendix A. There were two items (minor landscaping improvements at the front entrance and incomplete painting of administrative areas) that required adjustments but not retesting. The client verified these two adjustments on May 15, 2005. One item (food concession at the main entrance) was found to have unsatisfactorily slow access to the food preparation facility. KLSJ moved a number of fences and benches in the front of the concession booth and added an extra stove unit to improve pedestrian service. These amendments were retested on May 16, 2005.

Acceptance

Having approved the retest and improvements required as a result of the initial acceptance test, Carlington Aquatic Parks accepts the Ottawa–Carleton Water Park facility.

Carlington Aquatic Parks ____________________ **Dated** __________

KLSJ Consulting ____________________________ **Dated** __________

Appendix A: Acceptance Test Results

The Acceptance Test results are shown in Table 15.1.

Ottawa–Carleton Water Park, May 10, 2005

Grades:

A: Exceeds contract specification

B: Meets contract specification

C: Fails to meet contract specification owing to minor deficiencies; requires correction

D: Fails to meet contract specification owing to major deficiencies; requires retest

Table 15.1: Acceptance Test Results

Serial	*Test Subject*	*Test Parameters*	*Test Method*	*Score*	*Remarks (mandatory for C and D)*
1	Slides	• Performance specifications	• Weighted dummy with instrumentation sent down each slide minimum of 10 times	B	• Within all parameters; no test failures; some smoothing required
		• Environmental test	• Water quality test	A	
		• Appearance	• Visual inspection	A	
		• Overall impression		A	
2	Amusement attractions	• Performance specifications	• As specified in individual manufacturer documents	A	
		• Environmental test	• Water pumps and machinery in good working order	A	• No fluid or hydraulic leaks
		• Appearance	• Visual inspection	A	
		• Overall impression		A	
3	Grounds	• Performance specifications	• Walk-around inspection; compare work to specifications and drawings	B	• Complete front walkway; some painting required
		• Environmental test	• Visual inspection for spills; soil tests if warranted; confirm spills have been remediated	B	• Tests were generally satisfactory, but some soil still needs to be removed
		• Appearance	• Visual inspection	C	• Considerable construction material still to clean up; contractors have been advised
		• Overall impression		B	

(Continued)

Table 15.1: Acceptance Test Results (*Continued*)

Serial	*Test Subject*	*Test Parameters*	*Test Method*	*Score*	*Remarks (mandatory for C and D)*
4	Safety systems	• Performance specifications	• Inspection against standards; test equipment at random	A	
		• Environmental test	• Inspect for leaks	A	
		• Appearance	• Visual inspection for obstructions	A	
		• Overall impression		A	
5	Food services and concessions	• Performance specifications	• As specified in individual manufacturer documents	D	• Main snack bar too slow; design improvements needed
		• Environmental test	• Water quality tests; grease traps and ventilation testing; public restroom testing	B	• Some vent modifications required
		• Appearance	• Visual inspection	A	
		• Overall impression		B	
6	Park operations staff	• Performance specifications	• Verify personnel documents for training in lifeguarding, water safety, cardiopulmonary resuscitation	A	
		• Environmental test	• Verify personnel documents for training in WHMIS,	B	• Some training still to be completed

		• Appearance	chemical storage, herbicide and pesticide use	A	
		• Overall impression	• Observation during training	B	
7	Transportation	• Performance specifications	• Visual inspection; test drive	A	
		• Environmental test	• Emissions test; visual inspection for leaks	B	• Some heavy equipment must have emissions systems upgraded
		• Appearance	• Visual inspection	A	
		• Overall impression		A	
8	Sanitation	• Performance specifications	• Test evacuation in liters per minute; test all pumps and drainage systems; test maximum pounds per square inch	A	
		• Environmental test	• Leakage inspection	B	• Some minor intermittent leaks
		• Appearance	• Verify systems not within public view	A	
		• Overall impression		B	

Closing Phase Documents

16

Post-Project Report

Post-Project Report: Discussion

The Post-Project Report, sometimes called the Project Completion Report, is the final document drafted by the project team. This report formally documents project performance against planned project goals and objectives. The purpose of the report is twofold: First, it is the final report to the client on how the project unfolded; second, it is an internal document to the project team that is to capture what went right and where improvements could have been made, in other words, lessons learned.

In drafting this report, there is a natural tendency to highlight the good areas of the project and downplay the aspects that did not go as well as they could or should have. If the project was a great success, some measure of self-congratulation on the part of the project team is well earned, but do not get carried away: Save the celebrations for the wrap-up party. Remember always that someday you or others in your company may do a similar type of project and an honest, factual Post-Project Report may keep you from repeating a costly error.

The report is relatively straightforward to draft. There should be nothing new in it. Rather, the report is a compendium of previously reported issues and problems, with lessons learned added. Everything, good and bad, should have been revealed in the regular project status reports (see Chapter 11) drafted throughout the project. This report should be relatively short. The actual length naturally will vary with the complexity of the project, but the aim is to keep it to a "readable length." Complex issues and lengthy financial data can be included as appendixes or, if possible, highlighted with reference to the main project files on the matter.

Post-Project Report Outline

Executive Summary

This is an optional section that summarizes the key points of the report. It should provide sufficient detail for the reader to understand the main issues addressed in the report. At a minimum, it should have brief comments on how the project fared against the triple constraints of cost, quality, and time. Normally, an executive summary would be only one or two pages.

Introduction

The Post-Project Report should be a stand-alone document so that at a later date anyone looking for lessons learned on the project can get an appreciation of the project as a whole. This section should provide a little history on the project, its objectives, and in particular what its constraints were in terms of cost, quality (requirements and performance), and schedule. Much of this information can be culled directly from the Project Concept (Chapter 1), the Project Charter (Chapter 3), and the Project Plan (Chapter 5).

Project Management Strategy

This section highlights the project management strategy as well as the structure of the project management team, the intended relationship between the client, and the project team, and a brief description of subcontracting or third-party arrangements, if any. Much of this information can be obtained from the Project Plan (Chapter 5) and the Project Charter (Chapter 3). The key part of this section is an honest assessment of how well the intended project management strategy worked and recommendations for any improvements that could be made in a similar future project.

Project Performance

Describe how the project fared with respect to schedule, cost, and quality. A blow-by-blow description of the project is neither necessary nor desired. Instead, provide a high-level overview of the key points in each of the three subject areas. Referencing the Work Breakdown Structure (WBS) (a high-level version can be attached as an appendix) is a straightforward method of highlighting the actual versus the planned project schedule. Similarly, for the cost section, minimize the number of figures in the main body of the report. A one- or two-page cost summary should be included as an appendix. Earned Value charts (see Chapter 10 in the Guide to the Project Management Body of Knowledge) can provide a useful visual presentation of how the project tracked against the planned schedule and the budgeted cost. These charts could be incorporated into the text of the report or included as an appendix, whichever is more appropriate. The final section on quality should be a short description of any problems that occurred and, most important, should state whether the final product met client expectations.

Evaluation of Plans and Approaches

This section of the report highlights the adequacy of the four key plans developed in the early stages of the project: Communications Plan (Chapter 6), Risk Management Plan (Chapter 7), Quality Management Plan (Chapter 8), and Procurement Plan (Chapter 9). Again, it is not necessary to comment on each of the plans at length; instead, rather summarize the key issues. What were the strengths and shortcomings of each of the plans? How could they be improved next time? If one or more of the plans was found to be particularly weak, more detailed comments about improvements could be attached as an appendix.

Risk Management

The purpose of this section is not so much to comment on the risk plan itself but to discuss some of the key issues that arose during the project and how they were managed. Specific comments should be made about whether the issues were identified as a risk, whether there was a mitigation strategy in place, and whether that strategy was effective. The main point is not to place blame but to capture the lessons learned to avoid similar pitfalls next time around. Only the main issues that directly affected the schedule, cost, and quality of the final product need be covered. Again, be brief: Focus on the lesson to be learned rather than providing a long, drawn-out explanation of what happened. The project status reports (see Chapter 11) will provide the bulk of the information for this section. It may be useful to refer to the specific status report so that subsequent readers of the Post-Project Report can find more detail if they need it.

Successes and Failures

This section continues the lessons-learned theme. One or two critical successes and any failures should be highlighted and explained. Be honest;

chances are, the reader (client or project team member) is already familiar with the main points of the issue. What is important is that everyone learn from his or her successes and failures. Tone is important here. The more negative issues should be drawn in a factual manner without attaching blame. However, it would not be inappropriate to recognize an individual or group of individuals for a job particularly well done.

Reusability

This is a key part of the report both for the client and for the project team, particularly if a similar project is likely to be proposed in the future. Almost any aspect of the project could be highlighted as being applicable to other projects. This could include an especially strong plan such as the procurement plan, how an issue was handled, or a process that was developed that could be applicable to a wider range of projects.

Recommendations and Conclusions

This section is intended primarily for an internal audience. Specific recommendations and conclusions should be made as to whether the company should undertake projects of this nature in the future.

Post-Project Report: Example

KLSJ Consulting

July 12, 2005

Post-Project Report

Ottawa–Carleton

Water Park

14 Palsen St., Ottawa, ON, Canada, K2G 2V8

Contents

Introduction

General Background

Dan Milks, president and CEO of Carlington Aquatic Parks of Ottawa, identified the market need and potential benefits of a water park in the National Capital Region in early 1999. He envisioned a family-oriented water and amusement facility that would provide a recreational outlet for visitors and local residents in a city lacking this type of outdoor recreational attraction. Mr. Milks issued a Project Concept document in September 2002 and began the process of marketing his idea to local government and investors. The concept was based on a spring 2005 opening for a water recreation facility in the Ottawa–Carleton region costing about Can$12 million with a further $4 million available for future development.

KLSJ was contracted to manage the project through to completion of the facility and then turn over the running of the Water Park to a management team that would be responsible for the ongoing operations of the Water Park, essentially a turnkey project.

Overall Objectives

The overall objective of the project as outlined in the Project Plan of October 2002 was the successful design and construction of the Ottawa–Carleton Water Park facility on behalf of Carlington Aquatic Parks. The Water Park was to be completed and ready for a scheduled opening of May 21, 2005, at a total cost of Can$12,450,000. This was a Class B estimate with a range of +25% to −10%. The intention of the owners (Carlington Aquatic) was to have a top-quality recreational facility with the following criteria:

- A site within the western portion of Ottawa–Carleton
- A capacity of 7000 guests per day
- Unique attractions, including an outdoor wave pool, an artificial river, and water slides
- A parking area adjacent to the Water Park capable of accommodating 1000 cars
- Service by local roads, bus transportation links, and bike trails

Overall Project Constraints

Facing the triple constraint of time, cost, and quality and given a fixed target date for the opening of the park and the low probability that extra funding could be located on short notice, it was recognized that any project-related problems probably would result in a decrease in the number of attractions or their quality. Carlington Aquatic Parks agreed to establish a management reserve of Can$500,000 (approximately 4% of the estimated project cost).

Project Management Strategy

General Structure

The intention of Carlington Aquatic Parks was that all responsibility for the management of the development and construction activities be contracted to KLSJ. Carlington was to remain actively involved in the marketing of the project to political offices, community groups, potential investors, and future customers. Carlington also retained approval authority for all strategic aspects of the project. In that role Carlington approved financial plans and major expenditures (above Can$10,000), investors, architectural and engineering designs, and major contractors, as well as marketing activities.

Although KLSJ acted as overall Project Manager, specific aspects of project execution were contracted out to firms selected by KLSJ and approved by Carlington. Functions contracted to outside experts included the following:

- Environmental assessment (including environmental, architectural, and traffic studies)
- Engineering site services study
- Engineering design and construction
- Marketing

Project Team

The Project Team consisted of five core members for the duration of the project, plus three additional persons required to coordinate activities during the construction portion (Phase 3, Execution). The basic organization of the team is shown in Figure 16.1.

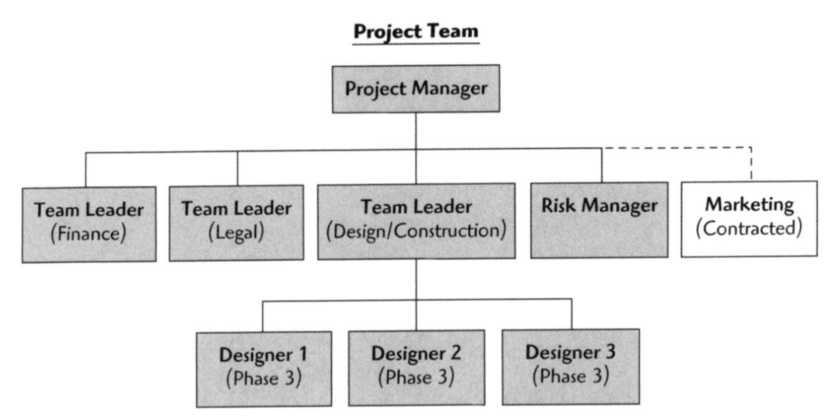

Figure 16.1 Project Team.

Roles and responsibilities for the individual team members are as provided in the Project Plan.

Comments on Project Management Structure

The overall project management structure functioned very well. The team was a manageable size, providing essential core expertise. However, as it would have been cost-prohibitive for the entire team to remain with the project full-time, resources were dedicated to the project on an as-required basis. The Project Manager and the Team Leader (Design and Construction) provided the necessary continuity. The financial, legal, and risk management personnel were supplied from KLSJ internal sources, whereas marketing expertise was contracted out. This arrangement provided the necessary flexibility to see the project through and enabled the team to react to changing requirements and priorities in a responsive manner. This "matrix" structure, however, required extra coordination and planning on the part of the Project Manager to ensure that resources were available as well as flexibility on behalf of the individual project members to adjust timetables to ever-changing requirements.

Lines of communication, both within and external to the project, were clear. The Project Manager was in direct communication with the client on an ongoing basis. Therefore, no major stumbling blocks were experienced in this regard. The client's requirements were clearly formulated, and the client was available to provide approvals as needed.

Project Performance

Schedule

As is evident on the WBS and Tracking Gantt schedule attached as Appendix A, the project tracked on or even slightly ahead of schedule for most of the early and middle portions. Although some individual activities, such as confirming investors and hiring the Operations Manager, took longer than anticipated, these delays did not affect the overall project schedule. Unfortunately, the combination of the late delivery of the water slides and the inclement weather in early spring 2005 forced a delay in the opening by four weeks to June 15. In accordance with the risk management criteria, this delay had a low to medium impact on the project.

Cost

The completed project was Can$311,650 (2.7%) over budget. A detailed cost breakdown showing cost variances is provided in the WBS and in Appendix B. A variety of individual activities contributed to cost overruns, including higher costs for the Water Park design, mitigation of an environmental problem, traffic and road upgrades, and a series of variances (positive and negative) in construction. The late delivery of the slides did not cause a significant cost increase, as the slide manufacturer paid for the majority of those costs. As can be seen in the Earned Value (EV) chart in Appendix C, EV tracked close to the actual for the entire project. Both EV and actual expenditures diverged from the budget baseline when the slides were not delivered as scheduled in September 2004. The gap was closed with the installation of the slides in spring 2005. The schedule performance (SP) and cost performance (CP) ratios tracked relatively close to 1 for most of the project. The spike in the SPI line in December 2002 is an anomaly caused by a rather large delivery being met at the end of a month rather than as scheduled early in the following month.

Quality

No significant quality compromises had to be made in the project. The small cost overrun did not adversely affect the number or quality of the planned attractions. The attractions that were delivered fully met the requirements of the client.

Evaluation of Plans and Approaches

Separate management plans were developed for risk, procurement, quality, and communications early in the project. Those plans proved invaluable in providing guidance and direction to a variety of project management staff. The following specific comments are made with respect to each of these areas.

Risk Management

Risk management was one of the most difficult aspects of the project, particularly in the Planning phase. Once site approval was obtained and funding was in place, the level of risk abated. The Risk Management Plan provided the essential framework to carry out this work. The Risk Management

Working Group provided a forum for discussing risk-related issues on a monthly basis. The highly structured approach to risk management was a key ingredient of success. "Risk Management," below, discusses specific risk issues in more detail.

Procurement Planning

The overall approach to procurement planning was sound. It was most useful to identify up front the areas that should be outsourced and the procedures for doing so. There was no difficulty contracting the consulting and construction firms, and suitable suppliers of materials were available. Unfortunately, the manufacturer of the water slides failed to deliver as promised, ultimately causing a delay in the overall project. This issue is discussed more fully in "Delay in the Delivery of the Slides," below.

Quality Management

A comprehensive Quality Management Plan provided a successful foundation for the quality program. KLSJ's experience in quality issues ensured a smooth approach to this area throughout the project. The key issue here was getting agreement from the owner-client on the relative importance of cost, quality, and schedule and where the trade-offs could be made. In the end, no significant quality trade-offs had to be made.

Communications

Communication, both internal and external, functioned very well. The client was readily available and took a keen interest in the project without unnecessarily encumbering progress.

Risk Management

Risk management proved to be a crucial aspect of the Water Park project. A comprehensive Risk Management Plan coupled with a functional Risk Management Working Group was imperative both to the early identification of problem areas and to reacting quickly to mitigate a risk as, or even before, it occurred. The following five major risk issues were dealt with over the course of the project.

Delay in Securing Investment Funding

When it became apparent that this activity was running behind schedule, extra resources were devoted to the task. The result was an overexpenditure of Can$12,000 on the activity (see attached Figure 16.2, ID 56). Although the activity ran six weeks over schedule, there was sufficient slack that it did not affect the overall project schedule.

Environmental Assessment Mitigation

The environmental assessment identified a problem that required rectification prior to the commencement of construction. An agreement eventually was reached with the regional government to split the cost 50:50 between the region and Carlington Aquatics even though the problem was clearly identified with the previous use of the site by the region. The cleanup (Figure 16.2, ID 142) cost the project Can$51,800, but the agreement with the region prevented a major slippage in schedule.

Delay in Hiring Suitable Operations Manager

This activity (see Figure 16.2, ID 87) took significantly longer than anticipated. Extra resources were devoted to the recruiting effort, and ultimately a well-qualified candidate was located and hired before the overall project schedule was affected. The cost overrun amounted to Can$14,000.

Delay in Delivery of the Slides

This item, in conjunction with adverse weather, was the key reason the project was delayed four weeks. Even though the slides were ordered well in advance and a reputable supplier was selected, the manufacturer failed to deliver the slides in time for the fall construction period (Figure 16.2, ID 105). The late delivery meant that some of the construction activities had to be shifted to the spring period immediately prior to the opening.

Inclement Weather Delays Construction

A late spring storm that deposited 30 cm of snow in the region caused a delay of two weeks in starting the installation of the slides. Additional delays because of the foul weather included extra cleanup of the site and repairs to attractions already installed. As the problem occurred relatively late in the project, there was insufficient time to react by crashing other activities to regain the lost time. This was recognized early, and the decision was made on April 1 to delay the opening date by four weeks. Delaying a decision at that point and hoping to catch up would have adversely affected advertising and promotion of the grand opening.

Successes and Failures

Key Successes

One of the key successes of the project was the Project Team structure. The team had the right balance of size and resources to remain flexible and responsive to changing priorities yet possessed sufficient expertise to resolve problems and keep the project on track. A judicious balance between utilization of in-house resources and contracting out certain functions was critical to success. This was particularly true in the planning phase that took the project through the bewildering maze of the various layers of government to obtain site approval and permission to proceed with construction.

Significant Failure

The key disappointment in the project obviously was the late delivery of the water slides. This was the major contributor to the four-week delay in the opening of the Water Park. There are a limited number of suppliers of the large water slides needed for this type of water park. Despite careful screening of the company and ordering of the slides in what was considered sufficient lead time, the slides were delivered too late for installation during the intended fall construction period. In selecting a supplier for a critical item such as this, it is essential to choose only the most reliable provider. Ideally, penalty clauses need to be built into the procurement contract for late delivery to ensure that the supplier has an incentive to deliver as promised.

Reusability

The WBS provides two areas that could be reusable in the planning of future projects: the process to gain approval for a project to be constructed on regional government property (essentially Phase 2, Planning) and the process that was used to prepare and hand over the Water Park to the operations management team (Phase 4, Handover to Operations). Gaining approval to construct on regional property is a complex and time-consuming process involving several levels of government and a myriad of special-interest and community groups that are concerned about the effect of the Water Park on their communities. These concerns had to be resolved prior to gaining approvals from the various government authorities. There is no ready guide on how to navigate this route, and the Project Team acquired considerable knowledge about these processes. Although individual projects are quite different, the framework used for this project would be a good foundation on which to build.

The second portion of the project that could prove to be of future value is the handover format that was used to turn the Water Park over to the ongoing management team in what was essentially a turnkey project. Turning over a product to a client in an efficient manner requires considerable planning and preparation, particularly in something as complex as a water park. The project team needs to ensure that this is carried out in a very structured and comprehensive manner.

Recommendations and Conclusions

Despite the end delay and slight cost overrun, the project can be considered a success from the perspective of KLSJ and that of the client. Labeling the project as high-risk from the onset focused the efforts of the entire team. Problems were not permitted to drag out but were dealt with expeditiously. The client clearly understood the issues and was a full member of the team.

It is recommended that KLSJ continue to accept turnkey projects of this type even if they are high-risk, assuming that the appropriate control mechanisms can be put in place at the start of the project.

Appendix A: Work Breakdown Structure and Schedule

The Work Breakdown Structure and Schedule is shown in Figure 16.2.

ID	Task Name	Start	Finish
1	**OC Water Park**	**'02 Aug 05**	**'05 May 20**
2	**Phase 1 - Concept**	**'02 Aug 05**	**'03 Mar 14**
27	**Phase 2 - Planning**	**'03 Feb 17**	**'04 Jan 16**
75	**Phase 3 - Execution**	**'04 Jan 19**	**'05 Jan 07**
76	**Project Management**	**'04 May 10**	**'04 Oct 29**
84	**Contract Management**	**'04 Jan 19**	**'04 Jul 30**
85	select parking contractor	'04 May 10	'04 Jun 04
86	develop operations plan	'04 May 10	'04 Jul 02
87	hire operations management team	'04 Jul 05	'04 Jul 30
88	select construction management firm	'04 Jan 19	'04 Feb 13
89	select design engineers	'04 Jan 19	'04 Feb 13
90	select water/sewage contractor	'04 Jul 05	'04 Jul 30
91	select road/traffic light contractor	'04 Jul 05	'04 Jul 30
92	select electrical transformer contractor	'04 Jul 05	'04 Jul 30
93	approve operations plan	'04 Jul 02	'04 Jul 02
94	**Financing**	**'04 Jul 05**	**'04 Sep 10**
99	**Construction**	**'04 Feb 16**	**'04 Oct 29**
124	**Marketing**	**'04 Jan 19**	**'05 Jan 07**
131	**Phase 4 - Turnover to Operations**	**'05 Feb 21**	**'05 Apr 22**
132	**Project Management**	**'05 Feb 21**	**'05 Apr 22**
139	**Contract Management**	**'05 Apr 11**	**'05 Apr 15**
141	**Construction**	**'05 Feb 21**	**'05 Apr 08**
142	conduct final site clean up	'05 Feb 21	'05 Mar 18
143	carry out modifications/repairs after trial	'05 Apr 04	'05 Apr 08
144	**Phase 5 - Closing**	**'05 Apr 18**	**'05 May 20**

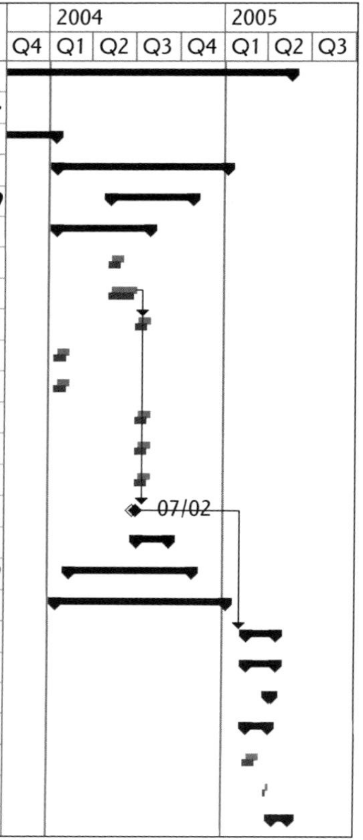

Figure 16.2 Work Breakdown Structure and Schedule.

Appendix B: Financial Summary

The Financial Summary is given in Table 16.1.

Table 16.1: Ottawa–Carleton Water Park Costs

Totals by Project Phase	Budgeted Costs				
	Fixed Costs	*Resource Costs*	*Total Costs*	*Actual Costs*	Variance *from Budget*
Phase 1: Concept	$25,000	$109,000	$134,000	$139,250	$5,250
Phase 2: Planning	$192,000	$185,550	$377,550	$415,600	$38,050
Phase 3: Execution	$10,584,000	$134,100	$10,718,100	$10,929,850	$211,750
Phase 4: Handover to Operations	$70,000	$28,500	$98,500	$155,100	$56,600
Phase 5: Closing	$0	$33,000	$33,000	$33,000	$0
Total	**$10,871,000**	**$490,150**	**$11,361,150**	**$11,672,800**	**$311,650**
Profit (10%)	$1,087,100		$1,087,100	$1,087,100	
Grand Total	**$11,958,100**	**$490,150**	**$12,448,250**	**$12,759,900**	

Totals by Project Category	*Budgeted Costs*				
	Fixed Costs	*Resource Costs*	*Total Costs*	*Actual Costs*	*Variance from Budget*
Project management	$432,000	$213,050	$645,050	$739,350	$94,300
Contract management	$0	$38,300	$38,300	$47,900	$9,600
Financing	$34,000	$83,200	$117,200	$126,200	$9,000
Political and legal	$0	$35,400	$35,400	$41,000	$5,600
Construction	$10,135,000	$91,100	$10,226,100	$10,419,250	$193,150
Marketing	$270,000	$29,100	$299,100	$299,100	$0
Total	**$10,871,000**	**$490,150**	**$11,361,150**	**$11,672,800**	**$311,650**
Profit (10%)	$1,087,100		$1,087,100	$1,087,100	
Grand Total	**$11,958,100**	**$490,150**	**$12,448,250**	**$12,759,900**	

Appendix C: Earned Value Chart

The Earned Value Chart is shown in Figure 16.3.

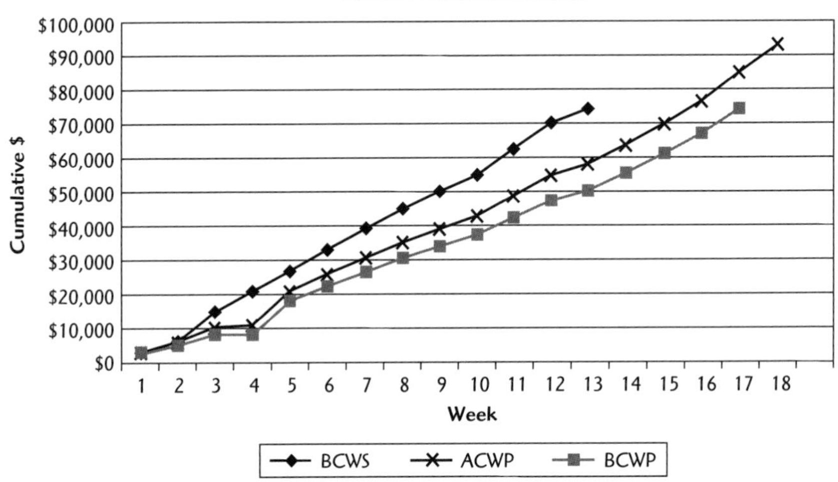

Figure 16.3 Earned Value Parameters.

17

Conclusion

Congratulations! You have made it to the end of the book. Before you put it down, the authors would like to leave you with some thoughts.

The documents outlined in the book do not have to be followed exactly in terms of format and content. Use each one as a framework: Start with the suggested format, add topics that you think are required, and delete ones that you feel are unnecessary. Unless you are running a construction project similar to this one, you most likely will have to change the documents significantly.

Similarly, if you are the manager of staff members who are required to produce these types of documents, allow them to change things: Add their improvements and modify the format to suit their specific needs. In the case of any changes, be sure to update the templates so that the modifications can be reused.

Finally, do not hesitate to e-mail any of the authors if you have any questions (see the Preface for e-mail addresses). The CD in the back of the book contains all the documents. Please use them as a starting point to customize your own documents. (If you use the text verbatim, please reference this book.) John Rakos is available to help you customize these templates or to provide training, seminars, or school lecturers on topics found in this book. Please see the Web site www.rakos.com for details on these services. The site also contains updates to the templates and examples and will include new templates as they are developed.

Good luck!

Appendix

About the CD-ROM

Introduction

This appendix provides you with information on the contents of the CD that accompanies this book. For the most current information, please refer to the ReadMe file located at the root of the CD.

System Requirements

- A computer with a processor running at 120 Mhz or faster
- At least 32 MB of total RAM installed on your computer; for best performance, we recommend at least 64 MB
- A CD-ROM drive

NOTE: Many popular word processing programs are capable of reading Microsoft Word files. However, users should be aware that a slight amount of formatting might be lost when using a program other than Microsoft Word.

Using the CD with Windows

To install the items from the CD to your hard drive, follow these steps:

1. Insert the CD into your computer's CD-ROM drive.
2. The CD-ROM interface will appear. The interface provides a simple point-and-click way to explore the contents of the CD.

If the opening screen of the CD-ROM does not appear automatically, follow these steps to access the CD:

1. Click the Start button on the left end of the taskbar and then choose Run from the menu that pops up.
2. In the dialog box that appears, type ***d*:\setup.exe.** (If your CD-ROM drive is not drive d, fill in the appropriate letter in place of *d*.) This brings up the CD Interface described in the preceding set of steps.

What's on the CD

The following sections provide a summary of the software and other materials you'll find on the CD.

Any material from the book, including forms, slides, and lesson plans if available, are in the folder named "Content."

Content

The book is divided into four sections, based upon the major project phases defined by the Project Management Institute's A Guide to the Project Management Body of Knowledge(PMBOK® Guide). These phases are Initiation, Planning, Execution/Control and Closing. Each of these phases produces one or more documents. Each section of the book is subdivided into chapters, one chapter dedicated to one document type. Therefore these documents appear in the same chronological order as produced during the implementation of the project. The CD found at the back of the book contains all of the document examples as Microsoft Word™ files. The reader is welcome to use these as a framework to develop his or her document.

Applications

The following applications are on the CD:

Adobe Reader

Adobe Reader is a freeware application for viewing files in the Adobe Portable Document format.

Word Viewer

Microsoft Word Viewer is a freeware viewer that allows you to view, but not edit, most Microsoft Word files. Certain features of Microsoft Word documents may not display as expected from within Word Viewer.

OpenOffice.org

OpenOffice.org is a free multi-platform office productivity suite. It is similar to Microsoft Office or Lotus SmartSuite, but OpenOffice.org is absolutely free. It includes word processing, spreadsheet, presentation, and drawing applications that enable you to create professional documents, newsletters, reports, and presentations. It supports most file formats of other office software. You should be able to edit and view any files created with other office solutions.

Shareware programs are fully functional, trial versions of copyrighted programs. If you like particular programs, register with their authors for a nominal fee and receive licenses, enhanced versions, and technical support.

Freeware programs are copyrighted games, applications, and utilities that are free for personal use. Unlike shareware, these programs do not require a fee or provide technical support.

GNU software is governed by its own license, which is included inside the folder of the GNU product. See the GNU license for more details.

Trial, demo, or evaluation versions are usually limited either by time or functionality (such as being unable to save projects). Some trial versions are very sensitive to system date changes. If you alter your computer's date, the programs will "time out" and no longer be functional.

Customer Care

If you have trouble with the CD-ROM, please call the Wiley Product Technical Support phone number at (800) 762-2974. Outside the United States, call 1(317) 572-3994. You can also contact Wiley Product Technical Support at **http://www.wiley.com/techsupport.** John Wiley & Sons will provide technical support only for installation and other general quality control items. For technical support on the applications themselves, consult the program's vendor or author.

To place additional orders or to request information about other Wiley products, please call (877) 762-2974.

Index

CUSTOMER NOTE: IF THIS BOOK IS ACCOMPANIED BY SOFTWARE, PLEASE READ THE FOLLOWING BEFORE OPENING THE PACKAGE.

IMPLIED WARRANTIES OF MERCHANTABILITY AND FITNESS FOR A PARTICULAR PURPOSE, WITH RESPECT TO THE SOFTWARE, THE PROGRAMS, THE SOURCE CODE CONTAINED THEREIN, AND/OR THE TECHNIQUES DESCRIBED IN THIS BOOK. JWS DOES NOT WARRANT THAT THE FUNCTIONS CONTAINED IN THE SOFTWARE WILL MEET YOUR REQUIREMENTS OR THAT THE OPERATION OF THE SOFTWARE WILL BE ERROR FREE.

(c) This limited warranty gives you specific legal rights, and you may have other rights that vary from jurisdiction to jurisdiction.

6. Remedies.

(a) JWS' entire liability and your exclusive remedy for defects in materials and workmanship shall be limited to replacement of the Software Media, which may be returned to JWS with a copy of your receipt at the following address: Software Media Fulfillment Department, Attn.: The Practical Guide to Project Management Documentation, Wiley, 10475 Crosspoint Blvd., Indianapolis, IN 46256, or call 1-800-762-2974. Please allow four to six weeks for delivery. This Limited Warranty is void if failure of the Software Media has resulted from accident, abuse, or misapplication. Any replacement Software Media will be warranted for the remainder of the original warranty period or thirty (30) days, whichever is longer.

(b) In no event shall JWS or the author be liable for any damages whatsoever (including without limitation damages for loss of business profits, business interruption, loss of business information, or any other pecuniary loss) arising from the use of or inability to use the Book or the Software, even if JWS has been advised of the possibility of such damages.

(c) Because some jurisdictions do not allow the exclusion or limitation of liability for consequential or incidental damages, the above limitation or exclusion may not apply to you.

7. U.S. Government Restricted Rights. Use, duplication, or disclosure of the Software for or on behalf of the United States of America, its agencies and/or instrumentalities "U.S. Government" is subject to restrictions as stated in paragraph (c)(1)(ii) of the Rights in Technical Data and Computer Software clause of DFARS 252.227-7013, or subparagraphs (c) (1) and (2) of the Commercial Computer Software - Restricted Rights clause at FAR 52.227-19, and in similar clauses in the NASA FAR supplement, as applicable.

8. General. This Agreement constitutes the entire understanding of the parties and revokes and supersedes all prior agreements, oral or written, between them and may not be modified or amended except in a writing signed by both parties hereto that specifically refers to this Agreement. This Agreement shall take precedence over any other documents that may be in conflict herewith. If any one or more provisions contained in this Agreement are held by any court or tribunal to be invalid, illegal, or otherwise unenforceable, each and every other provision shall remain in full force and effect.